Wanderson Stoco
Andre de Lima
Hamilton Torrezan

Measurement methodology for coordinate measuring machines - CMM

Wanderson Stoco
Andre de Lima
Hamilton Torrezan

Measurement methodology for coordinate measuring machines - CMM

Influence of the number of measurement points on shape errors

ScienciaScripts

Imprint
Any brand names and product names mentioned in this book are subject to trademark, brand or patent protection and are trademarks or registered trademarks of their respective holders. The use of brand names, product names, common names, trade names, product descriptions etc. even without a particular marking in this work is in no way to be construed to mean that such names may be regarded as unrestricted in respect of trademark and brand protection legislation and could thus be used by anyone.

Cover image: www.ingimage.com

This book is a translation from the original published under ISBN 978-613-9-61592-6.

Publisher:
Sciencia Scripts
is a trademark of
Dodo Books Indian Ocean Ltd. and OmniScriptum S.R.L publishing group

120 High Road, East Finchley, London, N2 9ED, United Kingdom
Str. Armeneasca 28/1, office 1, Chisinau MD-2012, Republic of Moldova, Europe
Printed at: see last page
ISBN: 978-620-7-88367-7

Table of contents:

I dedicate this work to my wife Carolina Maria dos Santos Schiavi, my parents Roseli Rueda Rodrigues Stoco and Antonio Reinaldo Stoco, and my brother Everton Augusto Stoco.

SUMMARY

Measuring shape errors on coordinate measuring machines is a process that is widely used today due to the ease and speed with which the results of a given characteristic can be obtained. The aim of this work was to establish a rule for determining the minimum number of points to be touched when measuring circularity in various shapes. For the case study, measurements were carried out on a pepa manufactured exclusively for this work, where the initial objective was achieved through calculations and comparisons. Based on the results of the measurements, a rule was defined which consists of touching 1 point/mm of the calculated perimeter, taking into account the nominal diameter specified in the drawing. The use of the rule allows for greater confidence in the results obtained, as it establishes a standardization for circularity measurements at various diameters and a reduction in the cost of measurement, since the time spent on this process is considerably reduced (72.9% in the case studied) depending on the number of points chosen by the metrologist.

Keywords: *GD&T.* Shape tolerance. Circularity. Metrology. MMC. Measurement uncertainty. Normalized error.

Chapter 1

1. Introduction

According to Wandeck (2010), in the mid-1940s, it was believed that error was inevitable and that everything produced contained a certain amount of off-spec peas. The manufacturing process necessarily had to go through two fundamental stages: Manufacture and Control 100% of the parts, in order to discard any products that had characteristics outside the design specifications. The concept of quality level was used, whereby all production was allowed to contain a certain number of bad parts and the consequence was a loss of resources of around 30%.

At the same time, with the rapid growth in demand for products and services, there was a need to control the manufacturing process more effectively in order to avoid wasting resources on stages that did not add value to the product, such as: Selection and Individual Adjustments. This is how *Geometric Dimensioning and Tolerancing (GD&T)* came about, with the aim of being a type of tolerance that guarantees interchangeability between parts and improves communication between the design and manufacturing sectors.

Shape and position errors are routinely checked using calipers and functional devices, but there are cases where it is necessary to check not only the quality but also the quantity of geometric tolerances. This brings with it the need to obtain figures by measuring with equipment that serves this purpose, such as feeler and comparator clocks.

According to Klen (2000), the invention of the Coordinate Measuring Machine (CMM) in the mid-1960s facilitated the process of 7
control of manufactured parts, as it avoids the need to set up complex systems with devices and clocks to obtain the shape and position errors.

This work consists of combining the ease of obtaining shape errors in MMC with the reliability achieved by choosing the right number of points to be touched to construct the geometric elements using the machine's *software.*

1.1. Objective

Based on the premise of "productivity with quality", the aim is to study the importance of a good choice of the number of points to be touched in a pepa, in order to obtain a shape error using MMC.

1.2. Specific objective

The study has as its specific objective some factors to be taken into consideration, as described:

*ƒ*Present the history of *GD&T,* as well as its theoretical basis;

f Demonstrate some of the concepts involved in the working principle of a CMM;

f Quantify the measurement uncertainty involved in the study carried out;

J By calculating the normalized error, compare the results obtained from the choice of the number of points in order to determine the minimum number of points for measuring a shape error without affecting the required quality.

1.3. Hypothesis

Establish a minimum number of points to be touched in a circularity measurement, in order to obtain a result that is as reliable as possible.

1.4. Justification

When faced with the need to measure a shape error using a CMM, the metrologist usually asks himself: How many points need to be touched to determine this characteristic?

The study of the influence of the number of measuring points on obtaining shape errors in MMC adds value to the company by increasing the standardization of measuring methods and the reliability of the results obtained.

1.5. Research problem

Is it important to determine the minimum number of points to be touched when measuring a shape error using an MMC?

1.6. Research methodology

In order to produce a high quality piece of work, various research sources were used, such as bibliographies, theses, magazine articles and content from reliable internet sites. The following methodology was used to develop this work:

Step 1: In this first stage, data on the subject was analyzed, taking into account the author's area of activity, where the knowledge acquired can be used to make improvements in everyday life;

Step 2: The theoretical basis was acquired through a bibliographical study on the subject, using books, magazines, articles and content from reliable Internet sites, where all the information necessary for the proper development of this work was collected from these sources;

Step 3: With the theoretical basis obtained in the literature review, data was collected through measurements, studies, observations and calculations, where it was possible to work on this data;

Step 4: Finally, based on the previous three steps, a conclusion can be drawn for the study in question.

1.7 Structure of the work

In the first chapter of this graduate work, the following were presented: The introduction which demonstrates the fundamental aspects about the study, the main objective of the work, the specific objective, the hypothesis, the justification regarding the topic, the problem that generated the idea about the subject and the research methodology in a succinct way.

The second chapter contains all the theoretical background on the subject in question, where all the theoretical background on the subject worked on in the case study was obtained and was used as a source for the necessary clarifications with a view to a good practical application of the study.

The third chapter deals with the exploratory case study and it is in this chapter that the use of the theoretical framework in practice is demonstrated, using the data from the research and the study itself.

The fourth chapter presents the conclusion of the case study.

Finally, the fifth chapter lists all the bibliographies used in the research for this course conclusion.

Chapter 2

2. Theoretical background

Metrology is a science that encompasses several areas of study, including: Dimensional, electrical, physical, chemical, temperature, among others. Within each area covered by metrology, there are numerous ramifications that can be used according to specific needs.

This work focuses on metrology in the dimensional area, where subjects relating to measurements in MMC, measurements of *GD&T* characteristics, calculations of the uncertainties pertinent to the measurement process used and comparisons by calculating the normalized error were covered.

2.1. Metrology

"Metrology is defined as the science of measurement and its applications." (VIM, 2012, p. 16).

According to VIM (2012), metrology encompasses all the theoretical and practical aspects of measurement, whatever the measurement uncertainty and field of application.

For Lira (2009), the SI is the foundation of modern metrology. The name International System of Units (SI) comes from the French *Système International d'Unités*. The SI is used internationally by legal agreements, even in countries with their own measurement system, such as the United States, which uses the *U.S. Customary System*. However, units such as inch, foot, yard, pound, etc. are defined in terms of the SI base units (1 *in* = 0.0254 m, etc.).

The act of measuring, according to Albertazzi and Sousa (2008), is the experimental procedure by which the momentary value of a physical quantity (measurand) is determined as a multiple and/or a fraction of a unit, established by a standard and recognized internationally.

"From a technical point of view, medipao can be used to monitor, control and/or investigate physical processes or phenomena." (ALBERTAZZI and SOUSA, 2008, p. 6).

Some definitions are important for the correct interpretation of terms related to metrology. These definitions have been taken from the International Vocabulary of Metrology, VIM (2012):

f Measurement accuracy: Degree of agreement between a measured value and the true value of a measurand;

f Measurement accuracy: Degree of agreement between indications or measured values, obtained by repeated measurements, on the same or similar objects, under specified conditions;

f Sensitivity of a measuring system: Ratio between the variation of an indication of a measuring system and the corresponding variation of the value of the measured quantity;

f Resolution: The smallest variation in the measured quantity that causes a perceptible variation in the corresponding indication;

f Measurement error: Difference between the measured value of a quantity and a reference value.

The VIM (2012) provides a set of definitions and associated terms in Portuguese and associated terms in English, French and Spanish for a system of fundamental and general concepts used in metrology.

2.1.1 Measuring instruments

According to Lira (2009), equipment known as measuring instruments are used to measure certain quantities. These are devices used to take measurements individually or in association with one or more supplementary devices.

Figure 1 shows examples of conventional measuring instruments that are routinely used on factory floors.

Figure 1: Conventional equipment

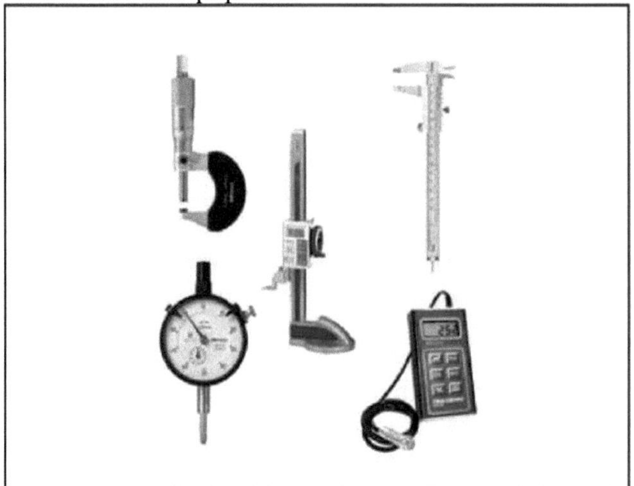

Source: MITUTOYO, 2012

Some equipment is more sophisticated and has a higher degree of complexity in its operation, as is the case with the MMC shown in figure 2.

Figure 2 Coordinate Measuring Machine - CMM

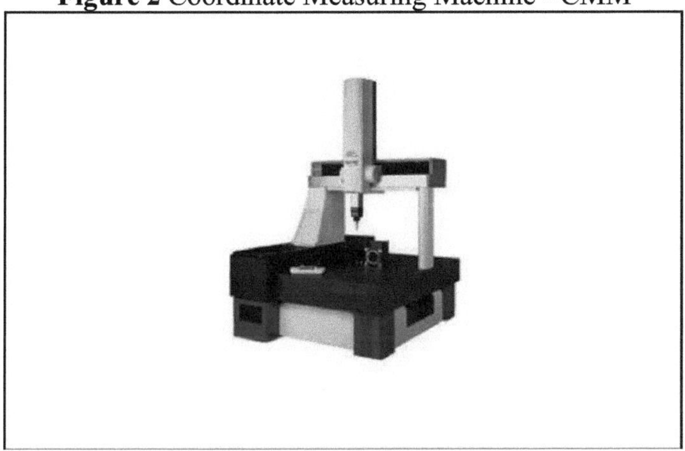

Source: MITUTOYO, 2012

CMMs are known for their flexibility in dimensional measurements. More detailed information on CMMs is described in section 2.1.7 "Coordinate Measuring Machines".

2.1.2 Calibration

Albertazzi and Sousa (2008) explain that measurement systems always present errors, without exception. To a greater or lesser degree, errors will always be present, even if they originate from within the measurement system or arise from the application of numerous quantities from external influences.

For the VIM (2012), calibration, in a first stage, is the operation that establishes, under specified conditions, the relationship between the values and measurement uncertainties provided by standards and the corresponding indications with the associated uncertainties. In a second step, it uses this information to establish a relationship in order to obtain a measurement result from an indication.

> Calibration of the instrument is necessary both to establish the correspondence between its indication and a standard value (and other metrological properties) and to determine corrections in the measurement process (LIRA, 2009, p. 134).

According to the VIM (2012), calibration should not be confused with the adjustment of a measurement system, often inappropriately referred to as "self-calibration", nor with calibration verification.

2.1.3. Calibration certificate

According to Albertazzi and Sousa (2008), a certificate must be generated to express the results of the calibration of a given measuring instrument. The calibration certificate is nothing more than a summary report containing important information about the calibration itself and its main results and relevant observations. When issued by a laboratory in the Brazilian Calibration Network, the certificate is an official document, issued on behalf of Inmetro.

According to Lira (2009), the calibration results must be documented in a calibration certificate, in a clear and objective manner, including the information requested by the clients, that relating to the method used and that necessary for interpreting the results themselves.

2.1.4 Brazilian Calibration Network

Albertazzi and Sousa (2008) point out that although the National Metrology Laboratory has the technical capacity to carry out calibrations in practically all quantities, it does not have the capacity to meet all of the country's demand. Due to the demand for calibrations to be carried out in the various industrial, commercial and scientific sectors, a country the size of Brazil cannot be served by just one laboratory for each quantity.

Albertazzi and Sousa (2008) state that Inmetro has the autonomy to officially accredit other laboratories in the country so that they can carry out officially recognized calibrations on its behalf. To do this, laboratories interested in accreditation must meet various requirements, demonstrating the technical and organizational capacity to carry out calibrations within the required levels of uncertainty. Laboratories accredited by Inmetro make up the Brazilian Calibration Network (RBC). The calibration certificates issued by the laboratories in the Brazilian Calibration Network are accompanied by the seal shown in figure 3.

Figure 3 Seal of the Brazilian Calibration Network (RBC)

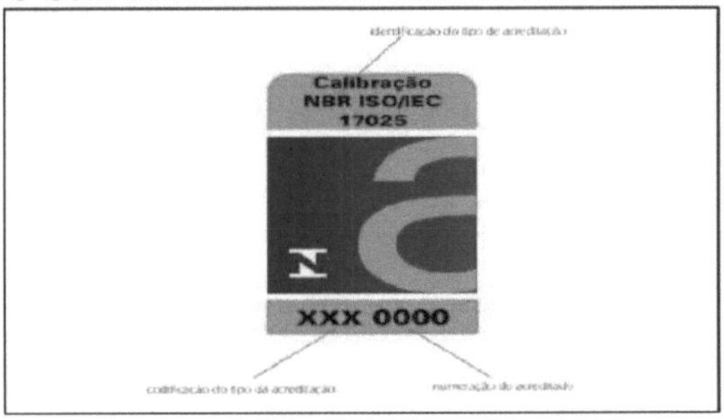

Source: NIE-CGCRE-009, 2012

The characteristics and dimensions of the CBR seal are governed by standard NIE-CGCRE-009 "Use of the Mark, Symbol and References to Accreditation" of 2012.

3.1.5. Normalized error

According to the guidance document Inmetro (2002), the normalized error (EN) technique is generally used as a means of comparing measurements to determine the performance of metrology laboratories and identify those with unacceptable measurement results. This is done by comparing each measurement result for a given characteristic made by several laboratories with the reference value that has been previously established.

If the results of a large group of participating laboratories differ significantly from the reference value, so that they clearly indicate the same trend, we can consider that the reference value has varied during the measurements.

"A convenient method for judging the quality of a measurement result is by calculating the normalized error in relation to the uncertainty." (INMETRO, 2002, p. 5). See equation 1.

$$En = \frac{resultado\ do\ laboratório - valor\ de\ referência}{((U_{LAB})^2 + (U_{REF})^2)^{\frac{1}{2}}} \quad (1)$$

Where:

U_{LAB} = Uncertainty of the Participating Laboratory;

U_{REF} = Uncertainty of the Reference Laboratory.

"*En* values less than unity indicate that the measurement is acceptable, i.e. the error is within the laboratory's measurement uncertainty." (INMETRO, 2002, p. 5).

2.1.6. GD&T

According to Strafacci (2009), GD&T stands for *Geometric Dimensioning and* Tolerancing.

Pastore ([200-]), in turn, describes GD&T as an acronym from the English *Geometrical Dimensioning and* Tolerances.

> *GD&T* is a powerful language used in the design of mechanical products. It is the link between the design of the product and its materialization. *GD&T* has a major influence on new product development, manufacturing, quality and the environment (WANDECK, 2010, p. 6).

Also according to Wandeck (2010), *GD&T* has some characteristics and among the main ones are:

S Language features that make it possible to specify the pepa's functional requirements clearly and precisely;

S Ensures the interchangeability of pepas;

S Allows Zero Defect assembly;

S Used by various technical standards such as: *ISO*, *DIN*, *ASME*, *JIS*;

S Required by the automotive industry through the QS-9000 standard.

Wandeck (2010) explains that the characteristics presented by *GD&T* provide: Shorter development times and costs, lower manufacturing costs and times and lower consumption of energy and raw materials, minimizing the impact on the environment.

2.1.6.1 Fundamental principle of GD&T

Wandeck (2010) reveals that Stanley Parker was the inventor of the *GD&T*. Parker was an engineer at a British Navy torpedo factory in Alexandria, Scotland".

According to Wandeck (2010), in 1940, when *GD&T was* created, there was a perception that error was inevitable, i.e. everything that was produced contained a quantity of off-spec peas. The production process at that time necessarily went through two stages: Manufacture and Control, in order to discard any parts that had characteristics outside the specifications predetermined in the project. In general, the concept of Quality Level was used, where all production was allowed to contain a certain amount of "bad" peas. The consequence of using this concept was a waste of resources of around 30%. Parker found that in several cases, products assembled with failed peps worked satisfactorily. As a result, it was concluded that many parts considered to be non-compliant were in fact good parts and what was wrong was the concept of a "bad" part proposed by the Cartesian system. By analyzing the tolerance zone, it was possible to understand how theoretically failed peppers could result in good products. Figure 4 has been dimensioned according to the concept of the Cartesian system, so it can be seen that the tolerance zone for the position of the hole is a square with a side of 0.2, within which the center of the hole can be freely located.

Figure 4 Dimensioned drawing according to the Cartesian system

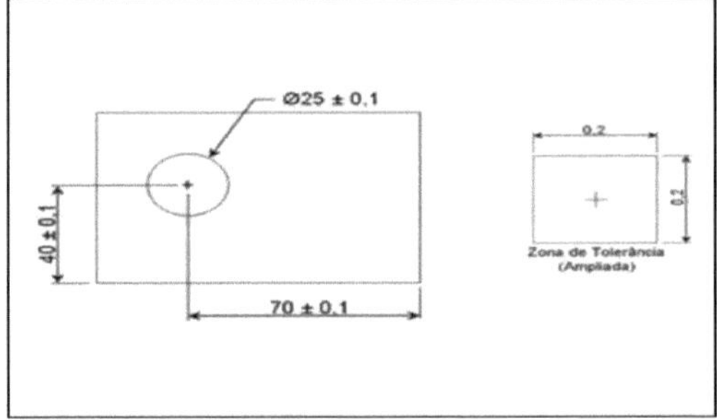

Source: WANDECK, 2010

According to Wandeck (2010), Parker's studies found that the center point of the tolerance zone is the perfect position, known as the true position. The ideal construction occurs when the center of the real hole agrees with the true position, represented by point 1 in figure 5. Point 2 is also a hole within the specifications. Point 3 is a "bad" hole because it is outside the tolerance zone, but point 3 is closer to the true position than point 2 and is still considered a non-conforming hole according to the Cartesian plane.

Figure 5 Tolerance zone according to the Cartesian system

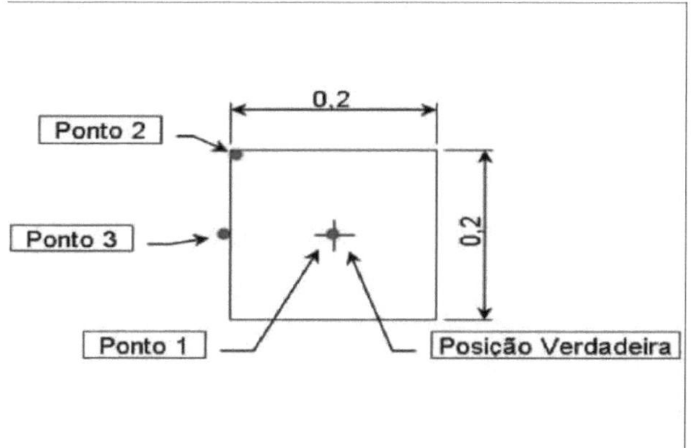

Source: WANDECK, 2010

From his experiments, Stanley Parker found that the critical characteristic for guaranteeing the interchangeability of products is the distance from the center of the actual hole to the true position. It was observed that point 3 is closer to the true position than point 2, i.e. if hole 2 is considered good, all the others located at the same distance from the true position should also be considered compliant holes. Parker therefore concluded that the square tolerance zone of the Cartesian system does not represent the functional assembly requirements of the products and in certain cases fails good parts (located in the hatched area of figure 6). Stanley Parker proposed a new format for the positioning tolerance zone, which changed from square to circular.

Figure 6: Circular tolerance zone

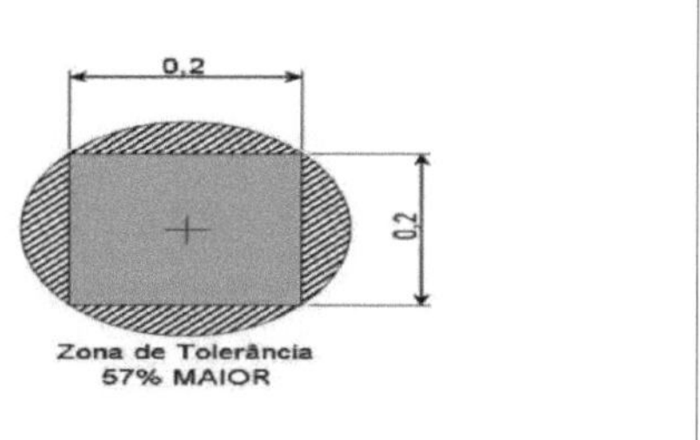

Source: WANDECK, 2010

It can be seen that the area of the circular zone is 57% larger than the area of the square zone and this very significant difference in the size of the tolerance field is directly reflected in the cost of the product.

"The circular tolerance zone is the fundamental concept of GD&T". (WANDECK, 2010, p. 11).

2.I.6.2 Types and symbols of geometric tolerances

The American standard ASME Y14.5 (2009) states that symbols are used to specify geometric tolerances, making it easier to visualize the type of tolerance to be considered, according to the functionality of the part to be tolerated. Table 1 describes all the geometric characteristics and their respective symbols.

Table 1: Geometric tolerances and their symbols

APPLICATION	TYPE OF TOLERANCE	CHARACTERISTIC	SYMBOL	SEE
INDIVIDUAL FEATURES	FORM	STRAIGHTNESS		5.4.1
		FLATNESS		54.2
		CRCULARITY		5.4.3
		CYLWORICITY		5.4.4
INDIVIDUAL OR RELATED FEATURES	PROFILE	PROFILE OF A LINE		62.1.2
		PROFILE OF A SURFACE		¿2.1.1
RELATED FEATURES	ORIENTATION	ANGULARITY		6.3.1
		PERP ENO CULAR TY		6 33
		PARALLELISM		6 3.2
	LOCATION	POSITION --		"
		CONCENTRICITY		7<4
		SYMME TRY		7.7.2
	RUNOUT	CIRCULAR RUNOUT		¿4.1
		TOTAL RUNOUT		"4.2

- Airheads may txs filed or <xx fined--May be r*a:ed or ixvtiaocd	3.3.11

Source: ASME Y14.5, 2009

There are additional instructions that are used to convey one or more pieces of information relating to the functional characteristics of the product. The additional symbols are shown in table 2.

Table 2: Additional symbols

TERM	SYMBOL	SEE:
AT MAXIMUM MATERIAL CONDITION (When applied to a tolerance value) AT MAXIMUM MATERIAL BOUNDARY (When applied to a datum reference)	Ⓜ	3.3.5
AT LEAST MATERIAL CONDITION (When applied to a tolerance value) AT LEAST MATERIAL BOUNDARY (When applied to a datum reference)	Ⓛ	3.3.5
TRANSLATION	▷	3.3.26
PROJECTED TOLERANCE ZONE	Ⓟ	3.3.6
FREE STATE	Ⓕ	3.3.20
TANGENT PLANE	Ⓣ	3.3.21
UNEQUALLY DISPOSED PROFILE	Ⓤ	3.3.22
INDEPENDENCE	Ⓘ	3.3.24
STATISTICAL TOLERANCE	⟨ST⟩	3.3T0
CONTINUOUS FEATURE	⟨CF⟩	3 3 23
DIAMETER	Ø	3.3.7
SPHERICAL DIAMETER	SØ	3.3.7
RADIUS	R	3.3.7
SPHERICAL RADIUS	SR	3.3.7
CONTROLLED RADIUS	CR	3.3.7

SQUARE	□	3.3.16
REFERENCE	( )	3.3.8
ARC LENGTH	⌒	3.3 9
DIMENSION ORIGIN	⌀—	3.3.17
BETWEEN	◄─►	3.3.11
ALL AROUND	⌀	3.3.19
ALL OVER	◎	3.3.25

Source: ASME Y14.5, 2009

The additional symbols play an important role in terms of guidance, both for the manufacturing process and for quality control.

J Straightness: "Straightness tolerance refers to the acceptable deviation of the shape of the element tolerated in the finished part, in relation to a straight line represented in the technical drawing". (PASTORE, [200-], p. 34). The symbol used to specify straightness is: u. Figure 7 exemplifies the application of the straightness tolerance, according to ISO 1101 (2004).

Figure 7: Application of the straightness tolerance

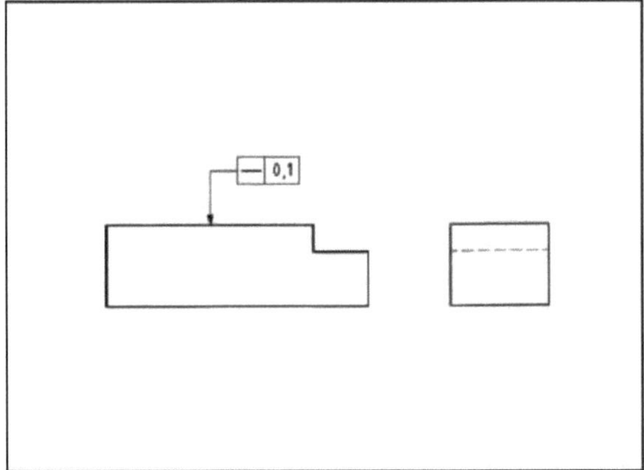

Source: ISO 1101, 2004

According to Mitutoyo ([200-]), a straightness tolerance is applied to the view in which the elements are controlled and are represented by a straight line.

J Flatness**:** "Flatness tolerance is the acceptable deviation in the shape of the tolerated element from the ideal flat shape". (PASTORE, [200-], p. 40). The symbol used to specify flatness is: C. Figure 8 shows an example of the application of the flatness tolerance, according to ISO 1101 (2004).

Figure 8: Application of flatness tolerance

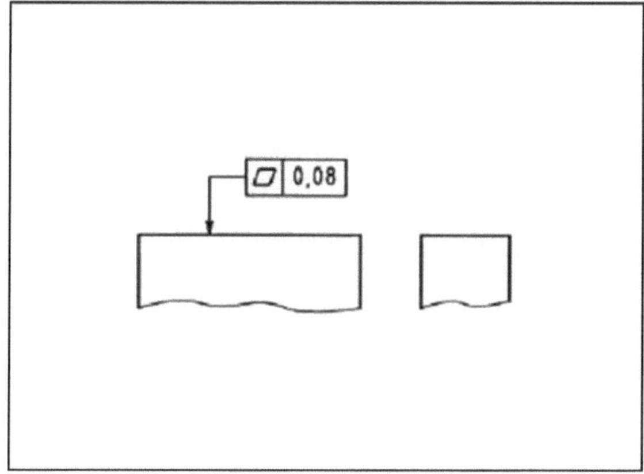

Source: ISO 1101, 2004

s **Circularity:** "Circularity tolerance corresponds to the deviation from the circular geometric shape that can be accepted without compromising the functionality of the pepa". (PASTORE, [200-], p. 44). The symbol used to specify circularity is: e. Figure 9 shows an example of the application of the circularity tolerance, according to ISO 1101 (2004).

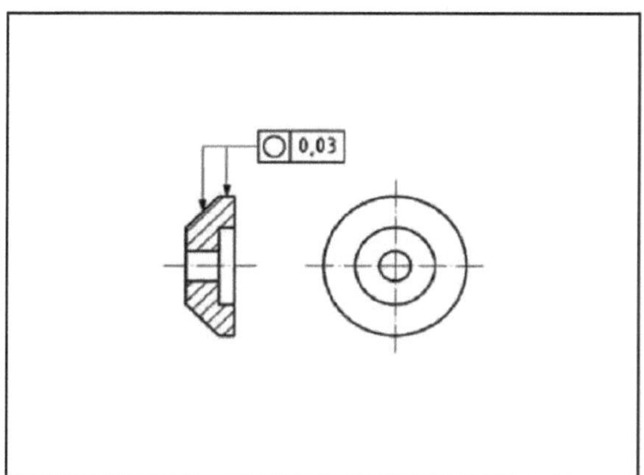

Figure 9 Application of the roundness tolerance
Source: ISO 1101, 2004

"This characteristic is mainly tolerated in conical and cylindrical parts." (PASTORE, [200-], p. 44).

J **Cylindricity:** "This is the acceptable deviation of the actual cylindrical surface in comparison with the ideal cylindrical surface represented in the drawing". (PASTORE, [200-], p. 48). The symbol used to specify cylindricity is: g. Figure 10 shows an example of the application of the cylindricity tolerance, according to ISO 1101 (2004).

Figure 10Application of cylindricity tolerance

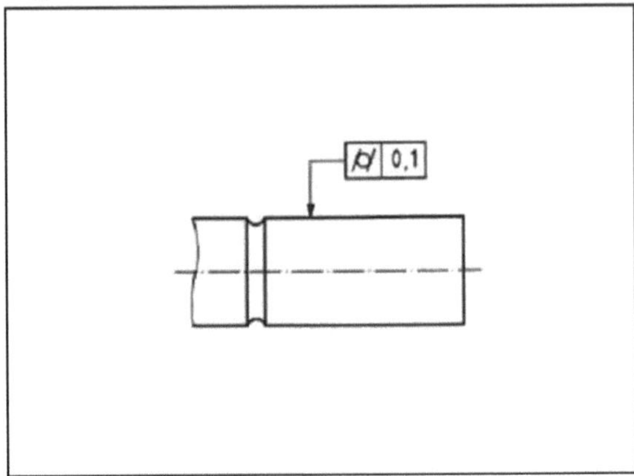

Source: ISO 1101, 2004

Mitutoyo ([200-]) comments that cylindricity tolerance is a composite of shapes including: Circularity, straightness and pendants of a cylindrical feature.

J **Line profile:** Pastore ([200-]) comments that in some cases, the accuracy of the irregular shapes of lines with profiles composed of radii and concordances may be essential for the functionality of the piece and that in order to guarantee this accuracy, it is necessary to specify the line profile tolerance. The symbol used to specify the line profile is: k. Figure 11 shows an example of the application of the line profile tolerance according to ISO 1101 (2004).

Figure 11 Application of line profile tolerance

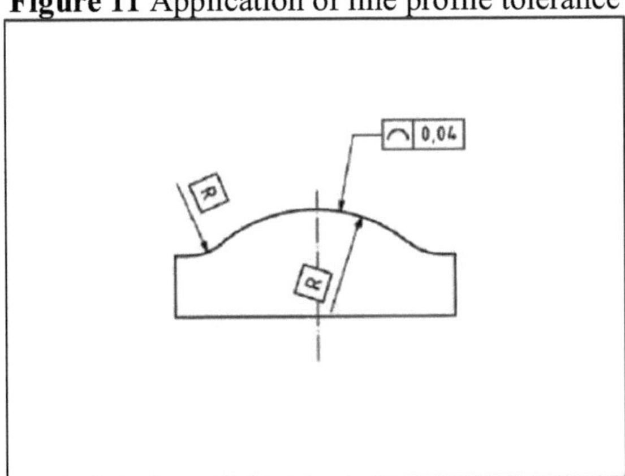

Source: ISO 1101, 2004

According to Pastore ([200-]), the line profile tolerance can also be applied to associated elements.

J **Surface profile:** According to Pastore ([200-]), the surfaces of parts can also have irregular profiles, made up of radii and matches, and when the accuracy of the irregular surface is a fundamental requirement for the functionality of the part, it is necessary to specify the surface profile tolerance. The symbol used to specify the line profile is: d. Figure 12 exemplifies the application of the surface profile tolerance, according to ISO 1101 (2004).

Figure 12 Application of surface profile tolerance

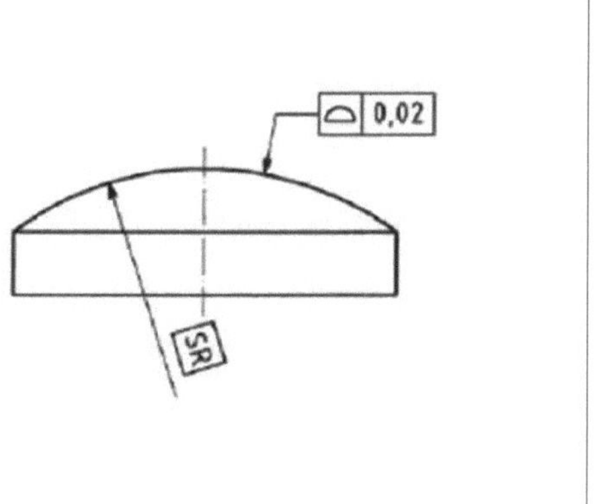

Source: ISO 1101, 2004

According to Pastore ([200-]), surface profile tolerance can also be applied to associated elements.

J Parallelism: According to ASME Y14.5 (2009), the parallelism tolerance is the condition of a surface or centerline where its points are equidistant to the *datum*. The symbol used to specify parallelism is: f .

According to ISO standard 1101 (2004), parallelism can be verified by considering two types of geometric elements: line and plane. It's worth remembering that a line can be the center line of a cylinder, a cone, etc. With the line and plane elements, there are also four possibilities of specification: line in relation to plane, plane in relation to line, line in relation to line and plane in relation to plane. Figure 13 shows the parallelism specification of line to plane, where the line is the axis of the toleranced cylinder and plane B is the reference.

Figure 13 Parallelism of a line to a plane

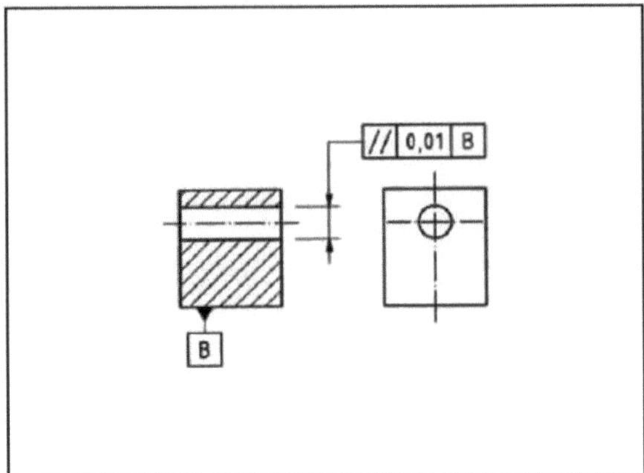

Source: ISO 1101, 2004

Figure 14 shows the parallelism specification of the plane with respect to the line, where the plane is the surface of the toleranced element and the line is the axis of the cylinder (*datum* C).

Figure 14 Parallelism of a plane to a line

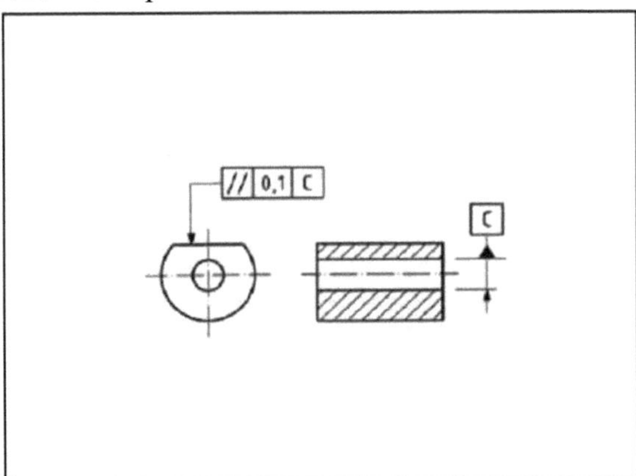

Source: ISO 1101, 2004

Figure 15 shows the specification of line parallelism with respect to the line, where one of the lines is the axis of the tolerated cylinder and the other line is the axis of the reference cylinder *(datum* A).

Figure 15: Line to line parallelism

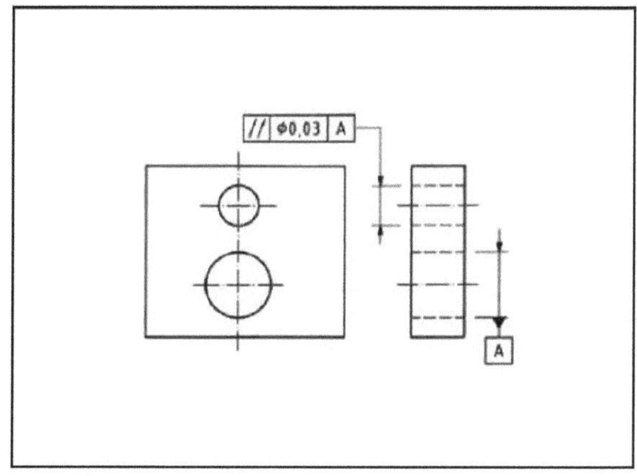

Source: ISO 1101, 2004

And finally, figure 16 specifies the parallelism of plane with respect to plane, where one of the planes is the tolerated surface and the other plane is the reference surface (*datum* D).

Figure 16 Parallelism of a plane to a plane

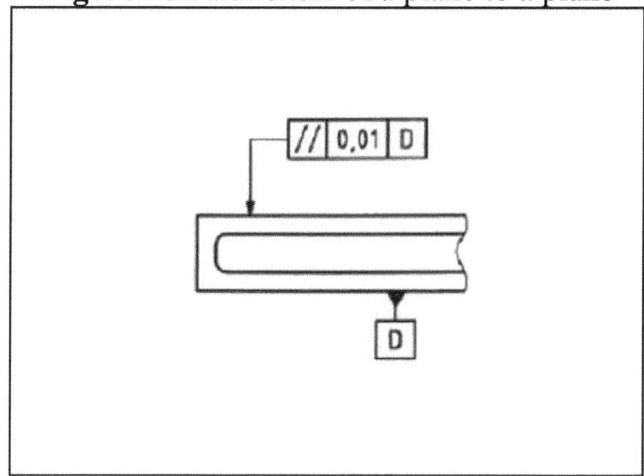

Source: ISO 1101, 2004

According to Mitutoyo ([200-]), parallelism tolerance is the condition of a surface or central plane, equidistant at all its points from a given plane, or an axis, equidistant along its length from one or more given planes to a given axis.

> **Perpendicularity:** According to ASME Y14.5 (2009), the perpendicularity tolerance is the condition where the toleranced element and the *datum* form an angle of 90°. The symbol used to specify perpendicularity is: b .

According to ISO 1101 (2004), like parallelism, perpendicularity can be checked by considering two types of geometric element: line and plane. It's worth remembering that a line can be the centerline of a cylinder, a cone, etc. With the

elements line and plane, there are also four possibilities of specification: Line in relation to
plane, plane relative to line, line relative to line and plane relative to plane. Figure 17 shows the specification of perpendicularity of line to plane, where the line is the axis of the toleranced cylinder and plane A is the reference.

Figure 17 Perpendicularity of a line to a plane

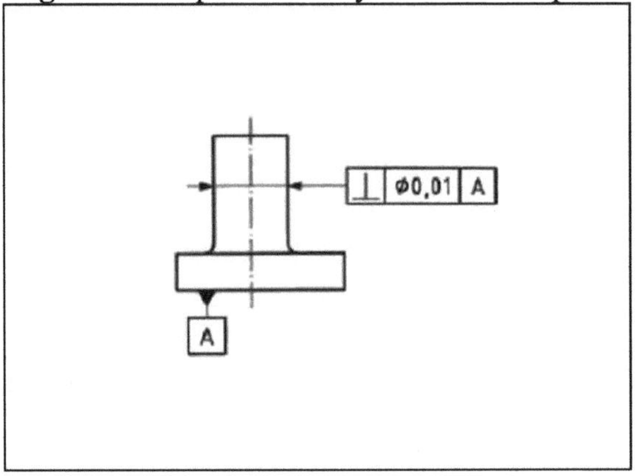

Source: ISO 1101, 2004

Figure 18 shows the specification of the perpendicularity of the plane to the line, where the plane is the surface of the toleranced element and the line is the axis of the cylinder (*datum* A).

Figure 18 Perpendicularity of a plane to a line

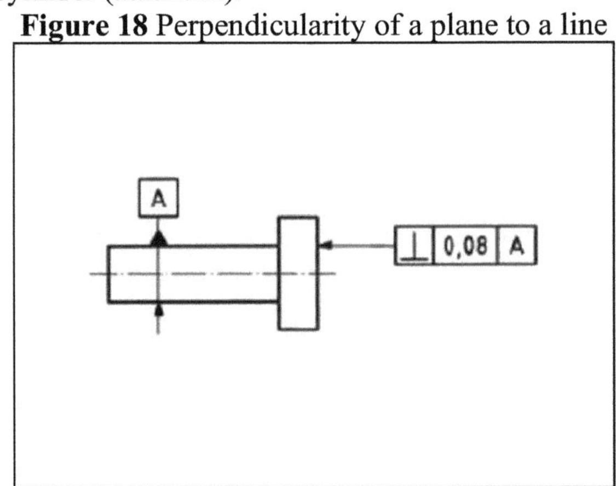

Source: ISO 1101, 2004

Figure 19 shows the specification of line perpendicularity to line, where one of the lines is the axis of the tolerated cylinder and the other line is the axis of the reference cylinder *(datum* A).

Figure 19 Perpendicularity of line to line

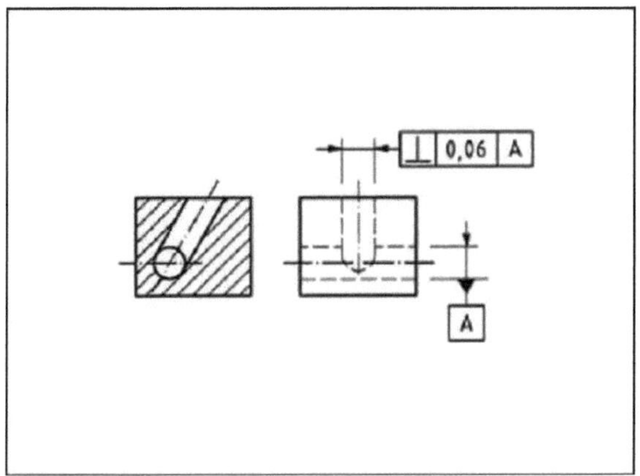

Source: ISO 1101, 2004

And finally, figure 20 specifies the perpendicularity of plane to plane, where one of the planes is the tolerated surface and the other plane is the reference surface (*datum* A).

Figure 20 Perpendicularity of a plane to a plane

Source: ISO 1101, 2004

According to Mitutoyo ([200-]), perpendicularity tolerance is the condition of a surface, central plane or axis at right angles to a given plane or axis.

J **Angularity:** According to ASME Y14.5 (2009), the angularity tolerance is the orientation of a surface or axis with respect to a *datum* plane or axis, used for angles of any value except 90°. The symbol used to specify angularity is: a .

According to ISO 1101 (2004), just like parallelism and perpendicularity, angularity can be checked by considering two types of geometric element: line and plane. It is worth remembering that a line can be the centerline of a cylinder, a cone, etc. With the elements line and plane you can also have four possibilities of specification: Line in relation to plane, plane in relation to line, line in relation to line and plane in relation to plane.

Figure 21 shows the specification of line angularity in relation to the plane, where the line is the axis of the toleranced cylinder and plane A is the reference.

Figure 21 Angularity of a line in relation to a plane

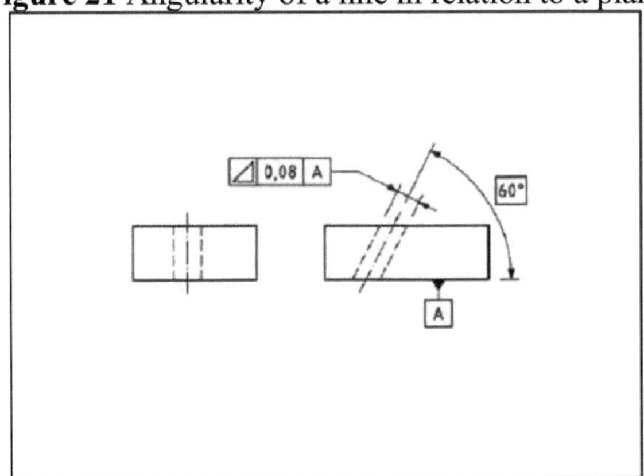

Source: ISO 1101, 2004

Figure 22 shows the specification of the angularity of the plane in relation to the line, where the plane is the surface of the toleranced element and the line is the axis of the cylinder (*datum* A).

Figure 22 Angularity of a plane in relation to a line

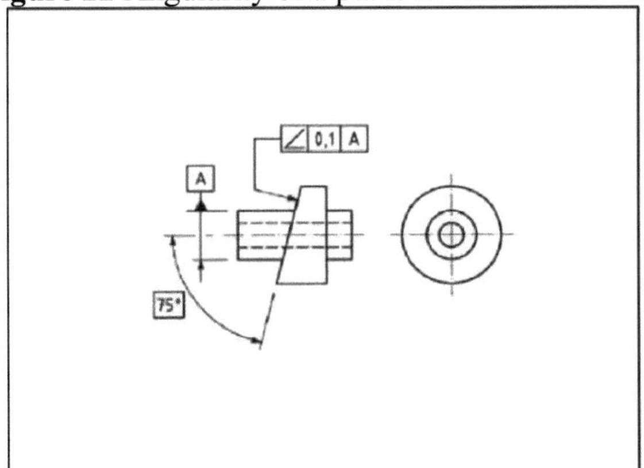

Source: ISO 1101, 2004

Figure 23 shows the specification of line angularity in relation to the line, where one of the lines is the axis of the tolerated cylinder and the other line is the axis of the reference cylinders *(datum* A-B).

Figure 23 Line angularity in relation to the line

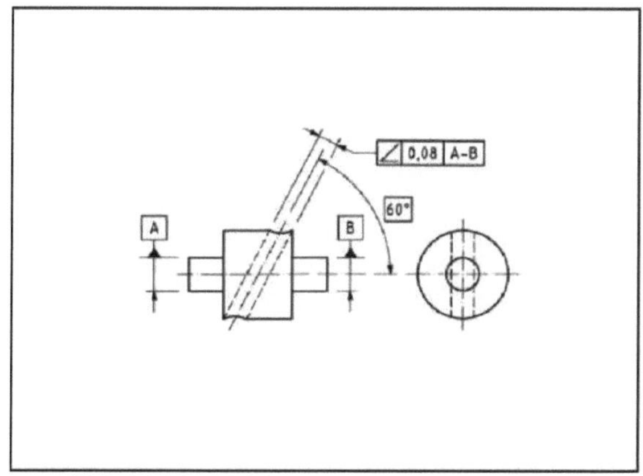

Source: ISO 1101, 2004

And finally, figure 24 specifies the angularity of the plane in relation to the plane, where one of the planes is the tolerated surface and the other plane is the reference surface (*datum* A).

Figure 24 Angularity of a plane in relation to a plane

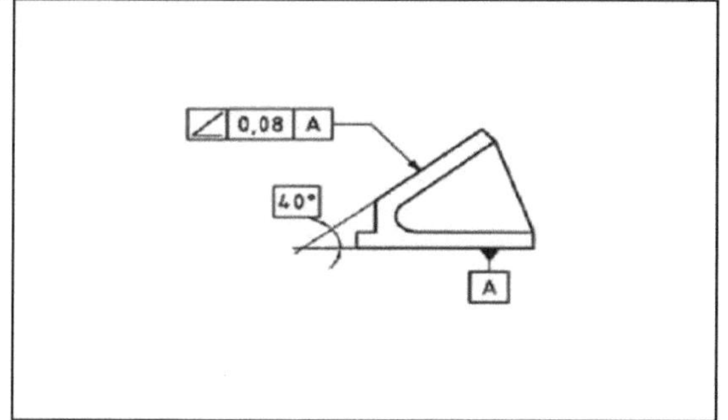

Source: ISO 1101, 2004

According to Mitutoyo ([200-]), angularity tolerance is the condition of a surface, central plane, or axis at a specified angle (other than 90°) from a given plane or axis.

> **True position:** The ASME Y14.5 (2009) standard specifies that the position tolerance is the permissible displacement that the elements can have without compromising the assembly. True position is the theoretically perfect position of an element. The symbol used to specify the true position is: j .

Figure 25 shows a hole that has a diametrical position tolerance of 0.08 mm in relation to C, A and B (*Datums:* Primary, Secondary and Tertiary respectively). In this case, the theoretically exact position is specified by the basic dimensions of 68 mm in relation to *datum* B and 100 mm in relation to *datum* A.

Figure 25 True position determined by basic dimensions

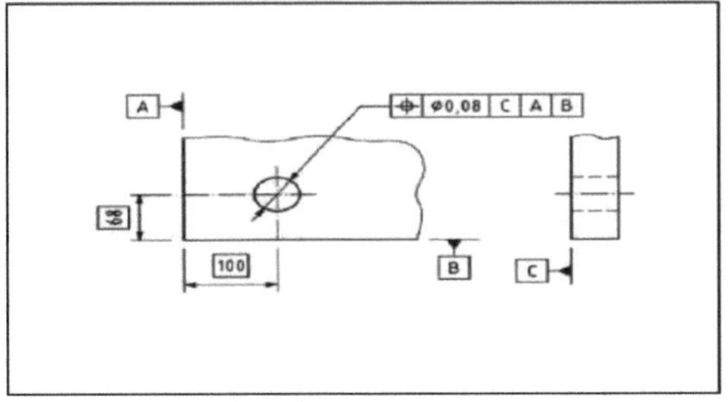

Source: ISO 1101, 2004

When specifying the position tolerance, the basic dimensions are used to define the theoretically exact position of the tolerated element.

> **Concentricity:** According to ASME Y14.5 (2009), the concentricity tolerance is the condition where the midpoints of the diametrical lines of the surfaces of revolution coincide with the centerline of the *datum*. The symbol used to specify concentricity is: r. Figure 26 shows an example of the application of the concentricity tolerance.

Figure 26 Application of concentricity tolerance

Source: ISO 1101, 2004

According to Mitutoyo ([200-]), the concentricity tolerance is the condition in which the midpoints of all the correspondingly located elements of two or more radially arranged features are congruent with the axis (or central plane) of a given feature.

> **Symmetry:** According to ASME Y14.5 (2009), the symmetry tolerance is the condition where the midpoints of the opposite elements are aligned with the centerline, or central plane, of the *datum*. The symbol used to specify symmetry is: ⌯ . Figure 27

gives an example of the application of symmetry tolerance.

Figure 27: Application of symmetry tolerance

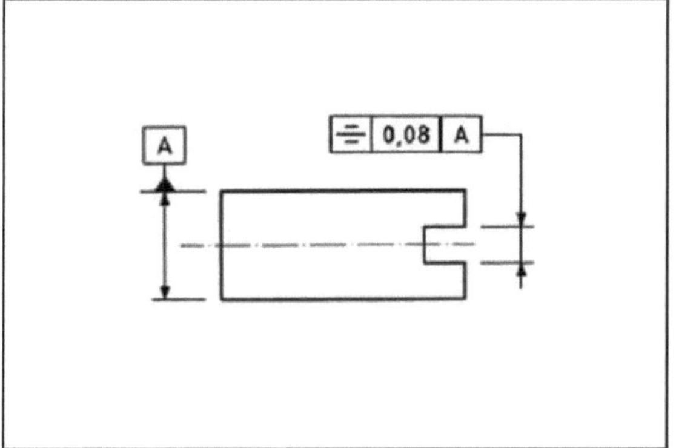

Source: ISO 1101, 2004

For Pastore ([200-]), the indication of symmetry in technical drawing presupposes the requirement of great rigor in the execution of the design. The symmetry tolerance specifies the limits within which symmetry errors can be accepted without compromising the functionality of the tolerated part.

> **Slamming:** According to SENAI (2007), the slamming tolerance represents the maximum permissible variation in the position associated with the shape of an element, observed when the part is completely rotated around a reference axis, i.e. when the part undergoes a complete rotation.

Depending on the checking position of the characteristic, you can have radial tapping or axial tapping. In radial tapping, the measurement is made with the dial indicator in a position perpendicular to the rotational axis of the workpiece, as shown in figure 28.

Figure 28: Radial beat

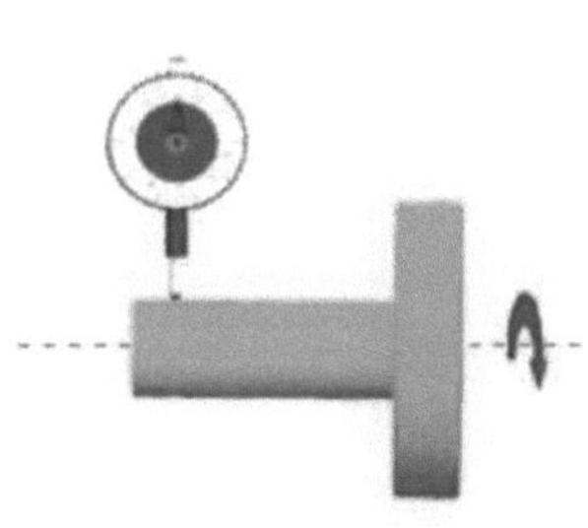

Source: WANDECK, 2010

For the axial beat, the measurement is made with the watch in a parallel position to the axis of rotation, as shown in figure 29.

Figure 29: Axial beating

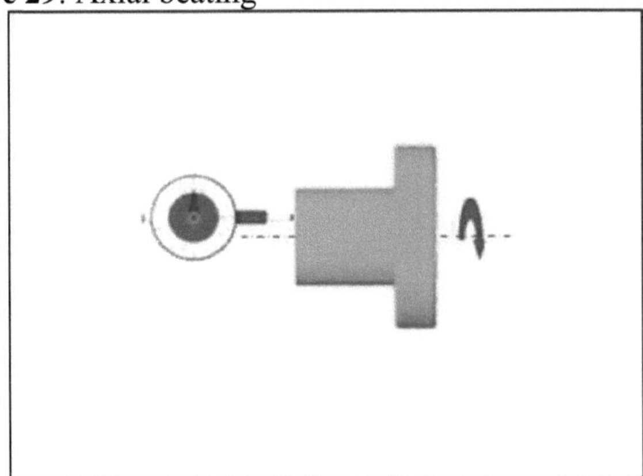

Source: WANDECK, 2010

According to Wandeck (2010), the tapping tolerance can also be used to control circumferences or surfaces. There is circular tapping, which is used to control the variation in the circumferences of straight sections, as shown in figure 30.

Figure 30: Circular beat

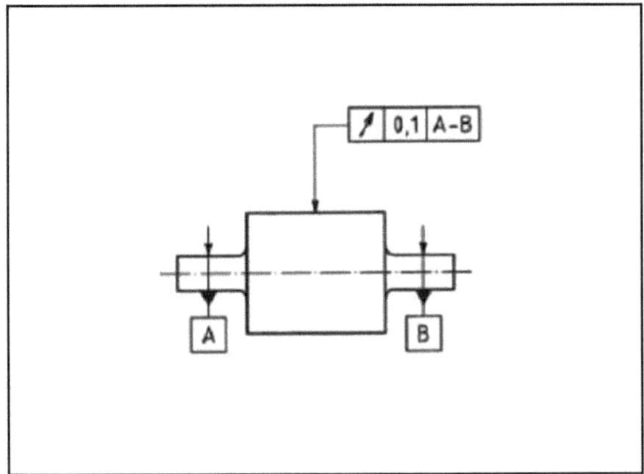

Source: ISO 1101, 2004

There is also the total beat which, according to Wandeck (2010), is used to control the variation of the surfaces of revolution. See figure 31.

Figure 31: Total beat

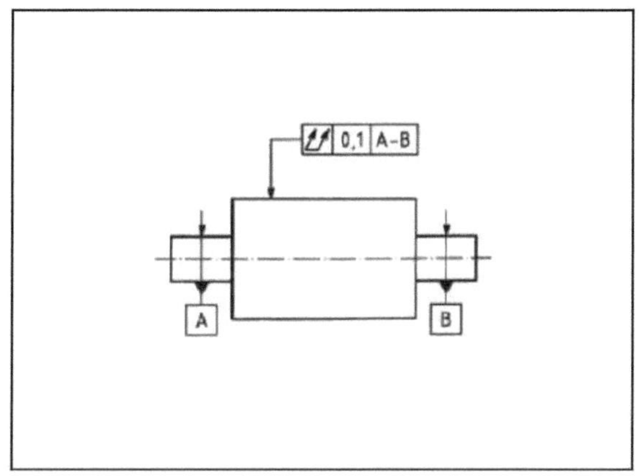

Source: ISO 1101, 2004

According to NBR 6409 (1997), the symbol used for specifying the circular beat is: and for the total beat is: .

2.1.6.3. *Datum*

According to ASME Y14.5 (2009), for *GD&T* language, *datum* is the element taken as a reference for establishing relationships imposed by geometric tolerances and for restricting degrees of freedom".

According to standard NBR 6409 (1997), the *datum* element or elements are identified by a framed capital letter connected to a filled or empty triangle, as shown by the blue arrows in figure 32.

Figure 32: *Datum* specification

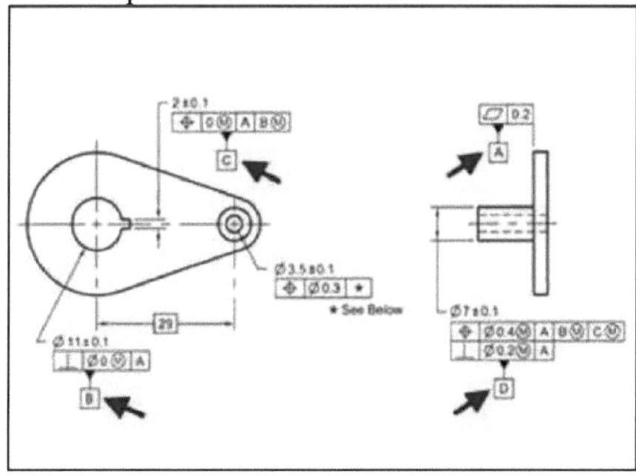

Source: ASME Y14.5, 2009

The same letter that represents the *datum* must be repeated in the tolerance table, as a way of establishing reference with the *datum* in question.

2.1.6.4. *Datum* on a flat surface

According to WANDECK (2010), if the *datum* element is a flat surface, the base of the triangle can be supported as shown in figure 33.

Figure 33: *Datum* on a flat surface

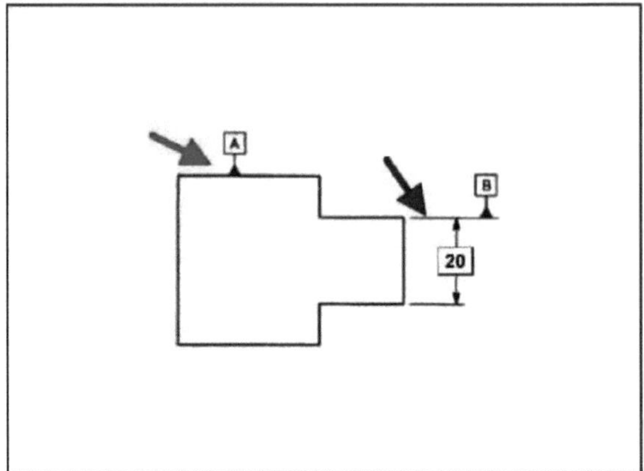

Source: WANDECK, 2010

The base of the triangle can remain:

> Directly on the surface, as indicated by the red arrow;

> On the extension line, as indicated by the blue arrow (in this case, the triangle must be in a clearly distinct place from the line containing the dimension or basic dimension, as shown by the *datum* B indication).

2.1.6.5. *Datum* on a cylindrical surface

Wandeck (2010) mentions that if the reference element is a cylindrical surface, its *datum* indication should be associated with the dimension of the cylindrical element. The *datum* can be shown on the drawing in various ways, as shown in figure 34.

Figure 34: *Datum* on a cylindrical surface

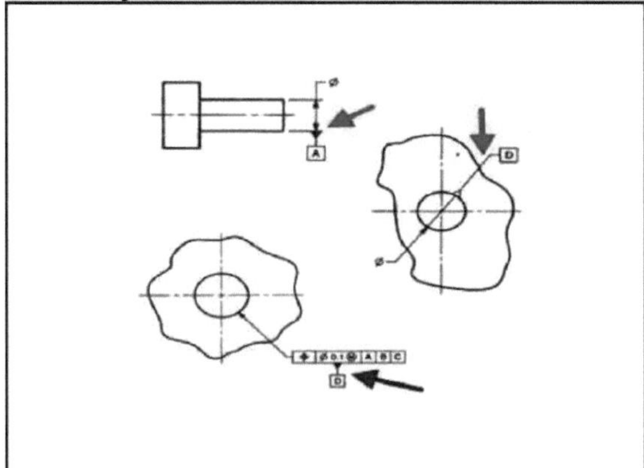

Source: WANDECK, 2010

> *Datum* on the extension of the line containing the dimension, as shown by the red arrows;
> *Datum* supported on the control board, as per the blue arrow.

2.1.6.6. *Datum* on a central plane

According to Wandeck (2010), if the *datum* element is the central plane of a slot or prism, the reference should be indicated as in figure 35.

Figure 35: *Datum* on a central plane

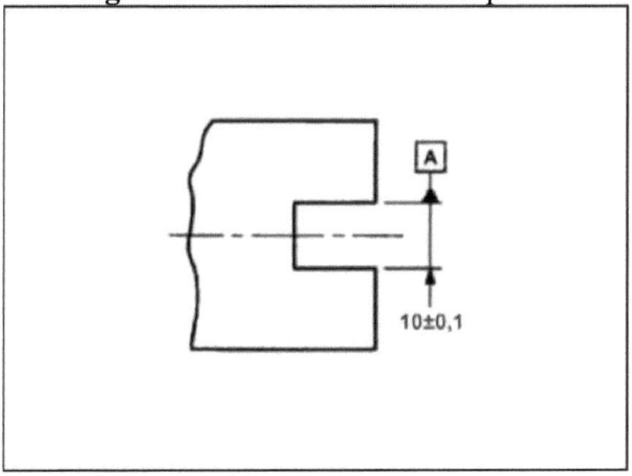

Source: WANDECK, 2010

"If the *datum* element is the central plane of a slot or prism, the symbol should be placed on the extension of the line containing the dimension." (WANDECK, 2010, p. 35).

2.1.6.7. Basic quota x reference quota

According to Mitutoyo ([200-]), the fundamental differences between basic dimensions and reference dimensions are:

> The basic dimension is the numerical value used to theoretically describe the exact size, profile or location of an element or reference location. From this dimension, the permissible variations are established through tolerances in other dimensions, notes or element control legends. The basic dimension has no tolerances and is shown inside a box (MITUTOYO, [200-], p. 9). See figure 36.

Figure 36: Basic quota

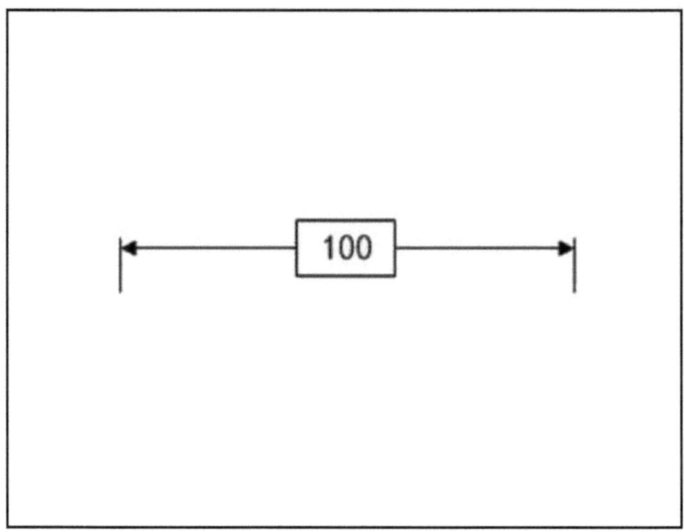

Source: MITUTOYO, [200-]

The reference dimension is usually a dimension with no tolerance for the purpose of auxiliary information. It is indicated in parentheses (MITUTOYO, [200-], p. 9). See figure 37.

Figure 37: Reference elevation

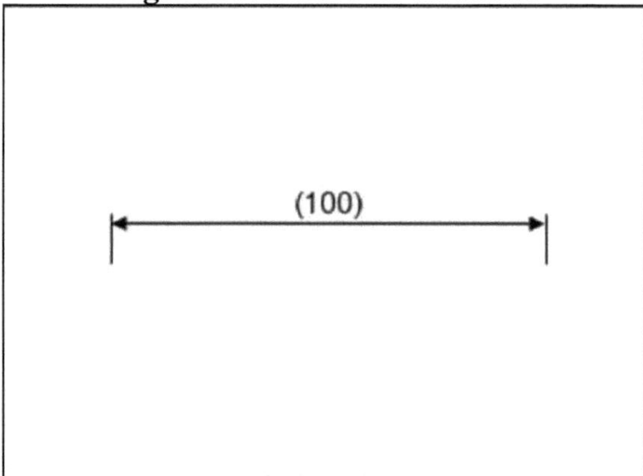

Source: MITUTOYO, [200-]

Basic dimensions are often confused with reference dimensions and this misinterpretation can lead to failures both in the manufacturing process and in the control of geometric tolerances.

2.1.6.8. Instruction box

According to Mitutoyo ([200-]), shape and position tolerances on drawings should be indicated in a rectangular box, divided into two or more parts according to the type of characteristic to be observed, as shown in figure 38.

Figure 38: Instruction box

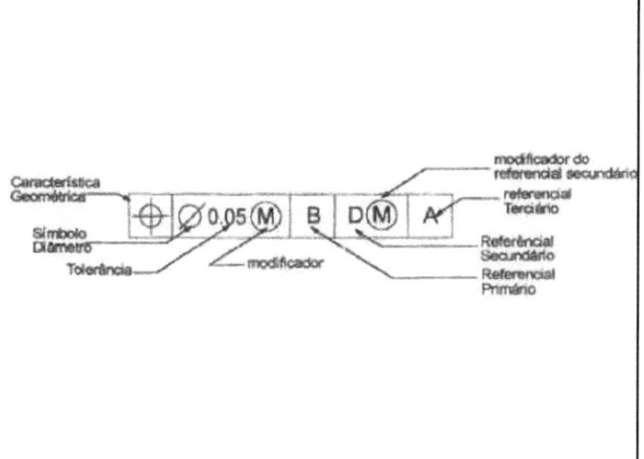

Source: MITUTOYO, [200-]

According to Mitutoyo ([200-]), the instruction box can also be called a control panel.

2.1.7 Coordinate Measuring Machine

According to PIEKAR (2006), a Coordinate Measuring Machine (CMM) is a piece of equipment that provides a representation of Cartesian, rectilinear and three-dimensional coordinate systems. The main function of this equipment is to compare the ideal geometry (prescribed in a drawing) of a given part with the geometry resulting from the measurement, such as size, shape, position, etc.

> In MMC, the geometry of a pepa is obtained by probing discrete points on the surface. Each measured point is expressed in the form of measured coordinates. However, it is not possible to evaluate the parameters of the pepa (e.g. diameter, distance, angle, etc.) directly from the coordinates of the measured points. Therefore, an analytical model of the pepa is used to evaluate these parameters. The model usually consists of ideal geometric shapes, also called substitute geometric elements. These elements are determined by applying best-fit algorithms to the measured points. The fitting algorithms used must be appropriate for each specific measurement task, depending on the function of the sample. The measurement system that was implicitly defined earlier is the basis of coordinate measurement technology (PIEKAR, 2006, p. 8).

Also according to PIEKAR (2006), coordinate measuring machines basically involve the subsystems detailed in figure 39.

Figure 39: Fundamental subsystems of a CMM

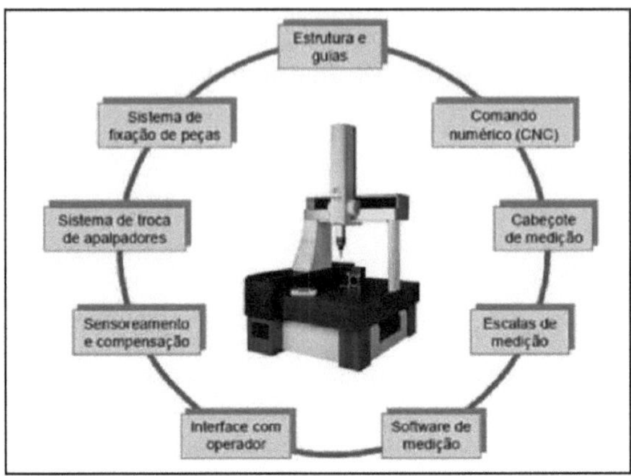

Source: PIEKAR, 2006

Klen (2000) states that for measuring parts with more complex characteristics, machines with coordinate measuring systems, due to their universality and high degree of automation, have established themselves as providers of quality assurance in industry.

2.1.8. Characteristics of coordinate measurement

According to Porath and Sousa ([200-]), coordinate measuring systems are currently part of a group of technologies that are already consolidated in dimensional quality assurance processes in various production sectors. The following are some of the advantages of coordinate measuring technology over conventional measuring systems:

- J **High flexibility:** Due to the characteristics of the measuring principle used in coordinate measuring, it can be applied with excellent performance in processes with small batches or unit samples, where the diversity of sample geometries and characteristics is great;
- J **High productivity:** With the use of CNC coordinate measuring machines, it is possible to work with large batches and high productivity, thus being able to use this measuring system in production processes with a high degree of automation;
- J **High computerization:** Easy integration with CAD, CAM and CAQ systems, providing excellent interaction between various sectors of the company;
- J **High precision**: In production environments where high precision is required, the use of coordinate measuring machines with measurement uncertainties in the order of ± 1 pm has become increasingly common;
- J **High reliability:** When used correctly and by properly trained personnel, this technology offers the same degree of reliability as any other type of dimensional control process.

According to Porath and Sousa ([200-]), in addition to its application in the dimensional control of products, the coordinate measuring system has proved to be a powerful tool in reverse engineering, through the precise digitization of free-form surfaces. In this way, coordinate measurement has been applied with excellent results at all stages of the production cycle of products, from design to serial production.

2.1.9. MMC *software*

According to BOSCH (1995), application *software* means the programs that control a CMM, as well as the computer and peripheral equipment, providing an interface between the measurement system and the operator. The *software* available today allows, among other things:

- *J* Integrate MMC into a computerized production environment (CAD/CAM/MMC);
- *J* Improve the accuracy of measurements;
- *J* Allow complete statistical analysis;
- *J* Provide real-time reports.

This work used COSMOS *software* with the main *Geopak-Win* module, version 2.4.R10 from the manufacturer Mitutoyo.

2.1.10. COSMOS *software*

According to Mitutoyo (2003), COSMOS is a versatile program that has advanced technology when it comes to dimensional measurement techniques. Using a computer, COSMOS makes MMC even more flexible thanks to the ability to use graphics, voice commands and the *Windows* operating system© . COSMOS has a package of modules consisting of: *Part Manager* (Program Manager), *Geopak-Win* (Geometric measurement and control of the CMM), *Geo-Scan* (Profile scanning and comparison to 2D), *Statpak-Win (*Statistical data analysis),

Transpak-Win (Converting *Geo-Scan patterns* to other patterns) and *3D-Tol-Win* (Comparing surfaces).

According to Mitutoyo (2003), *Geopak-Win* is the main COSMOS module. It is responsible for controlling the reading of CMM data and for carrying out geometric measurements on manual or CNC CMMs. Using this module, it is possible to measure, evaluate, calculate and document the geometric properties of a given part. The other modules are accessed from *Geopak-Win*.

2.1.11. Circularity

According to PASTORE ([200-]), the circularity tolerance is the acceptable deviation from the circular geometric shape that does not compromise the functionality of the piece.

Wandeck (2010) comments that circularity, being a shape tolerance, is not associated with dimension, location or orientation and also does not use *datums*. Circularity can be applied to cylindrical, conical or spherical surfaces.

Wandeck (2010) comments that, as shown in figure 40, the circularity tolerance zone is the space between two concentric circles orthogonal to the centerline, within which the surface under consideration must be located.

Figure 40: Tolerance zone for roundness

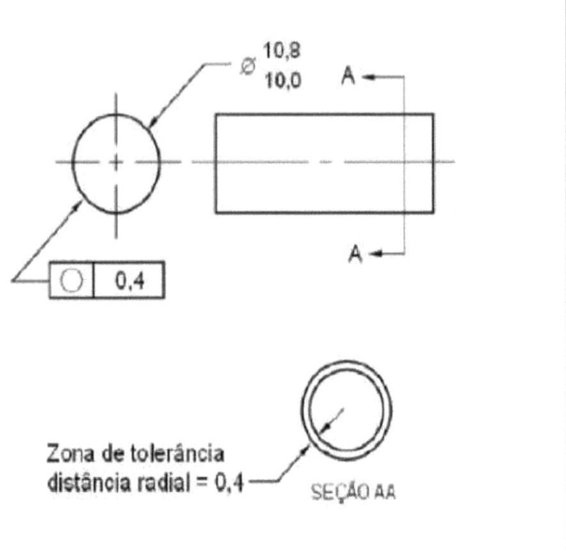

Source: WANDECK, 2010.

Circularity is measured using the COSMOS *software by* measuring a circle. According to Mitutoyo (2003), the circle element can be constructed in four ways:

J Gauss;
J Minimum circumscribed circle;
J Maximum circumscribed circle;
J Minimum zone element.

To carry out the measurements in this work, we used the Gauss circle element construction.

> Gauss: The program calculates the mean circle element. This element comprises points located in such a way that their distances on both sides (inner and outer) are approximately equal (or more precisely: The sum of the squares of the distances is as small as possible). (MITUTOYO, 2003, p. 65). See figure 41.

Figure 41: Circle calculated using the Gauss method

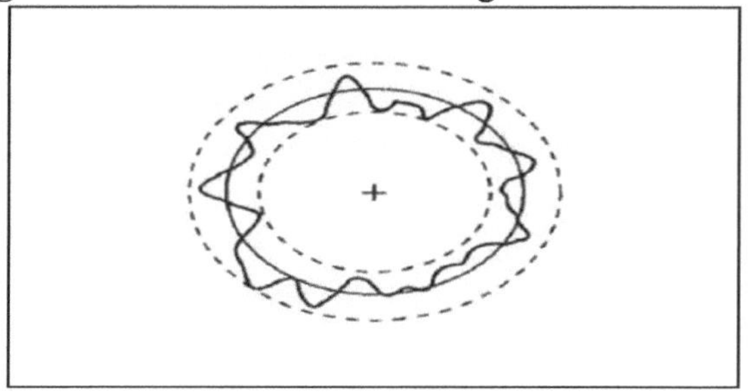

Source: MITUTOYO, 2003

From the measurement of the average circle, the circularity is obtained by considering all the points touched for the construction of the geometric element circle.

2.1.12. Measurement uncertainty

"Measurement uncertainty is a non-negative parameter that characterizes the dispersion of the values assigned to a measurand, based on the information used." (VIM, 2012, p. 24).

According to INMETRO (2003), the uncertainty of a measurement result reflects the lack of exact knowledge of the value of the measurand.

The result of a measurement must always be accompanied by an indication of the quality of the measurement. This indication must be easily understood and implemented. The concept of error and its analysis is old practice in metrology, but the concept of uncertainty as an attribute of reliability is relatively new (LIRA, 2009, p. 84).

According to Inmetro (2003), the result of a measurement, after correcting for recognized systematic effects, is only an estimate of the value of the measurand because of the uncertainty arising from random effects and the imperfect correction of the result for systematic effects.

2.1.13. Sources of uncertainty

According to Albertazzi and Sousa (2008), each factor that contributes to the uncertainty of the measurement process is called a source of uncertainty.

According to ABNT, Inmetro (2003), there are many possible sources of uncertainty that can influence a measurement, such as unrepresentative sampling, environmental conditions at the measurement site, personal bias error in the reading of analog instruments, inaccurate values of measurement standards, approximations and assumptions incorporated into the measurement method and procedure, variations in repeated measurements under apparently identical conditions, among others.

2.1.14. Evaluation of the measurement uncertainty of input estimates

ABNT, Inmetro (2003) explains that measurement uncertainty can be determined under repeatability conditions. The input quantities and their input estimates detected in the calibration process are taken from sources that can be measured, and are divided into two types: Type A and Type B.

2.1.14.1. Evaluation of type A standard uncertainty

According to ABNT, Inmetro (2003), in general, the best available estimate of the expectation or expected value ^q of a quantity q that varies randomly and for which n independent observations q_k were obtained under the same measurement conditions, is the arithmetic mean or average q of the n observations. See equation 2.

$$\bar{q} = \frac{1}{n}\sum_{i=1}^{n} qi \tag{2}$$

Where:

$\bar{q}$ – the average of the "n" nominations;
qi - individual value of each indication;
n - the number of repeated indications made.

According to Albertazzi and Sousa (2008), the standard uncertainty associated with one or more sources of uncertainty is often estimated from repeated measurements of the same measurand, i.e. n > 1. This is then a statistical procedure, routinely referred to as the "type A" procedure, which is nothing more than the experimental standard deviation. See equation 3.

$$u(I) = \sqrt{\frac{\sum_{k=1}^{n}(I_k - \bar{I})^2}{n-1}} \tag{3}$$

Where:

u(I) - standard uncertainty of the indications calculated from "n" repeated measurements;
Ik - k-th indication;
T - the average of the "n" indicators;
n - the number of repeated indications made.

2.1.14.2. Evaluation of type B standard uncertainty

Albertazzi and Sousa (2008) state that it is not always practical, or even feasible, to statistically obtain the standard uncertainty of a given source of uncertainty. Other available information must be used. This practice consists of using non-statistical procedures, also known as "Type B" procedures.

> For an estimate *xi* of an input quantity *Xi* that has not been obtained through repeated observations, the associated estimated variance u^2 *(xi)* or standard uncertainty *u(xi)* is evaluated by scientific judgment, based on all available

information about the possible variability of *Xi*. (ABNT, INMETRO, 2003, p. 11).

ABNT, Inmetro (2003) states that the set of information may include:

f Data from previous measurements;

f Experience or general knowledge of the behavior and properties of relevant materials and instruments;

f Manufacturer's specifications;

f Data provided in calibration certificates and other certificates;

f Uncertainties attributed to reference data taken from manuals.

"For convenience, $u^2(x_i)$ and $u(x_i)$ estimated in this way are sometimes referred to as, respectively, a Type B variance and a Type B standard uncertainty." (ABNT, INMETRO, 2003, p. 12).

2.1.15. Types of probability distribution

According to Albertazzi and Sousa (2008), the amplitude and type of distribution expected for the random component can be estimated or even deduced. The main probability distributions commonly used to model sources of uncertainty are (see figure 42):

A. Rectangular or uniform (Divider: $\sqrt{3}$);

B. Triangular (Divider: $\sqrt{6}$);

C. Normal or Gaussian (Divisor: Vñ or coverage factor k).

Figure 42: Most common probability distributions for sources of uncertainty

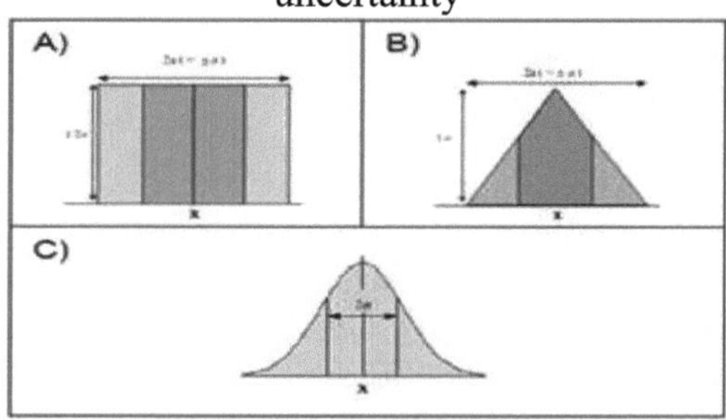

Source: QUAM, 2002

According to Albertazzi and Sousa (2008), the A distribution, rectangular or uniform, is used when there are equal chances of any value between -a and +a occurring or, in certain cases, for safety reasons, when there are doubts about which type of distribution to adopt. In the case of distribution B, the central values are more likely to occur. This is the type of distribution that naturally results from adding or subtracting two variables with the same rectangular or uniform probability distribution. The normal or Gaussian distribution is shown in box C. Since the curve of the normal distribution is asymptotic in relation to the horizontal axis, it should be made clear that the positions -a and +a are not the extremes of the curve, but correspond to the limits that cover approximately 95% of the area under the curve.

2.1.16. Evaluation of the measurement uncertainty of output estimates

According to ABNT, Inmetro (2003) the standard uncertainty is defined as being "a" standard deviation *u(xi),* for which we divide the value of the standard uncertainties (Type A and Type B) of the estimates of each input quantity by their respective divisor according to the type of distribution corresponding to the source of uncertainty. See equation 4.

$$u(x_i) = \frac{u_i}{d}$$

(4)

Where:

$u_{(xi)}$ – output uncertainty-pattern;

$u_{(i)}$ – uncertainty value for the input estimate;

d - divisor according to the type of probability.

2.1.17. Combined uncertainty (uc)

Albertazzi and Sousa (2008) explain that combined uncertainty is the standard deviation resulting from the combined application of the random components of all the sources of uncertainty affecting a measurement process. The results of the combined application of the various sources of uncertainty must be quantified by the combined uncertainty. See equation 5.

$$u_c^2 = u_1^2 + u_2^2 + u_3^2 + \cdots + u_n^2$$

(5)

Where:

U_c - combined uncertainty;

U_i - standard uncertainty of the i-th source of uncertainty.

2.1.18. Number of effective degrees of freedom (veff)

According to Albertazzi and Sousa (2008), the number of degrees of freedom corresponding to the combined uncertainty is called the number of effective degrees of freedom. This number is calculated from the values of the number of degrees of freedom and the standard uncertainty of each source of uncertainty using the *Welch-Satterthwaite* equation. See equation 6.

Where:

$$\frac{u_c^4}{v_{ef}} = \frac{u_1^4}{v_1} + \frac{u_2^4}{v_2} + \cdots + \frac{u_n^4}{v_n}$$

(6)

2.1.19. - number of effective degrees of freedom;
2.1.20. - combined uncertainty;
2.1.21. - uncertainty-pattern of the i-th source of uncertainty;
2.1.22. - number of degrees of freedom of the i-th source of uncertainty.

2.1.19. Scope factor (k)

According to VIM (2012), the span factor is defined as the number greater than "one" by which a combined standard uncertainty is multiplied to obtain an expanded measurement uncertainty. To obtain *k*, simply look it up in the *Student* table from the calculated v_{eff} value. For calibrations, laboratories usually use a probability of coverage of 95.45%, which in *Student*'s distribution results in table 3.

Table 3: *Student*'s table

	1	2	3	4	5	6	7	8	10	20	50	infinite
k	13,97	4.53	3,31	2,87	2.65	2.52	2,43	2,37	2,28	2.13	2,05	2,00

Source: LIRA, 2009

6.1.20. Expanded measurement uncertainty (U)

Combined uncertainty is defined by Albertazzi and Sousa (2008) as the value corresponding to the standard deviation of the joint action of all the sources of uncertainty. The range of values in which, with a 95.45% probability, the random error of the measurement process is expected to be found is obtained by multiplying the combined uncertainty (u_c) with the coverage factor (*k*). This gives a value equivalent to the repeatability of the combined action of all the sources of uncertainty, known as the expanded uncertainty, usually symbolized by the letter "*U*". See equation 7.

$$U = k.u_c \tag{7}$$

Where:

U - expanded uncertainty;
k - coverage factor;
uc - combined uncertainty.

Chapter 3

3. Case studies

In order to carry out the case study, a measurement strategy for the MMC was developed and experimented with, which allowed the results obtained in various circularity measurements to be compared on a given page, where, by varying the points touched to construct the geometric element circle, data can be obtained for the comparisons in question.

In all the comparisons in the study, the measurement uncertainties involved in the measurement process were taken into account, in order to achieve a probability of coverage of 95.45%, as mentioned by Albertazzi and Sousa (2008) in section 2.1.20.

The concepts of measurement and calculation based on the theoretical framework were applied, and at the end of the experiment the gain in terms of productivity, combined with quality, could be verified.

3.1. Materials and methods

Various pieces of equipment, accessories and materials were used to carry out the study:

S *Saddle* material: Apo SAE 1045 (Raw material: 03" x 35 mm);
S Conventional lathe, ROMI Tormax 30B;
S Machining center, ROMI Discovery D1000, Siemens 810D control;
S Flat grinding machine, MELLO;
S Drill 010 mm;
S Drill bit 05/8";
J *Vise* for clamping the workpiece in the machining center;
J Coordinate measuring machine, MITUTOYO Beyond Crysta 710, COSMOS *software* with *Geopak-Win* main module;
J 10 mm tip with 03 mm ruby ball;
J Standard sphere;
J Prismatic block for attaching the pepa to the MMC table;
J Thermohygrometer, Minipa;
J Digital stopwatch, Casio.

3.1.1. Manufacturing process

The entire process of designing the pepa, from concept to materialization, was based on the need to obtain a diameter with a relatively high circularity error, in order to enable comparisons between the different numbers of points and angular positions predefined for this study.

> **Manufacturing strategy**

Firstly, a design was drawn up with the 40 mm diameter as the main characteristic, allowing for a circularity error of up to 0.045 mm.

The pepa manufacturing process started with the drawing identified as TCC-001, revision 000.

> **Manufacturing procedure**

On the conventional lathe, the part was faced and the center hole was drilled, starting the machining process.

After marking the center, the first through hole was drilled with the 010 mm drill bit and then with the 05/8" drill bit.

The through internal turning was then carried out, leaving the measurement at 039.50 mm.

Then external turning, making a recess in the part measuring 070.50 mm x 14.50 mm long.

On the machining center, the workpiece was clamped in a vise.

Facing was then carried out on the reference surface.

This was done by interpolating the internal and external diameters to a height of 15 mm, leaving the measurements of 040.00 mm and 070.00 mm respectively, as shown in figure 43.

Figure 43: Machining the 40 mm and 70 mm diameters

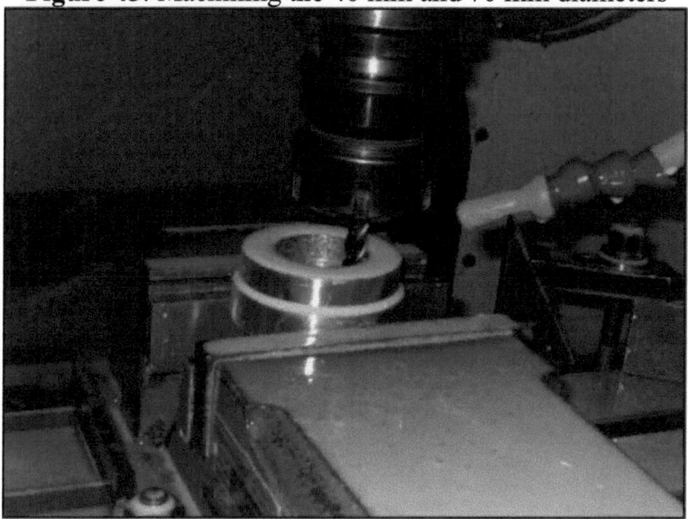

After the process on the machining center, the flat faces were ground, as shown in figure 44.

Figure 44: Flat face grinding

The entire process described in the "Manufacturing procedure" section resulted in an ideal skin for the study, thus achieving the initial objective considered when designing the skin.

3.1.2 Mediation process

Before the medipao began, a strategy was developed to compare different medipoes with different numbers of points measured.

This strategy was developed specifically for this job. In a routine measurement, the metrologist would be responsible for determining all the parameters to be taken into account in the measurement.

> **Measurement strategy**

First, the following items were determined: Starting angle, number of points to be probed and height at which the measurements were taken.

Each measurement was taken by starting the touches at three different angular positions, 0°, 30° and 60°, as shown in figure 45.

Figure 45: Angle positions for the first touch of each measurement

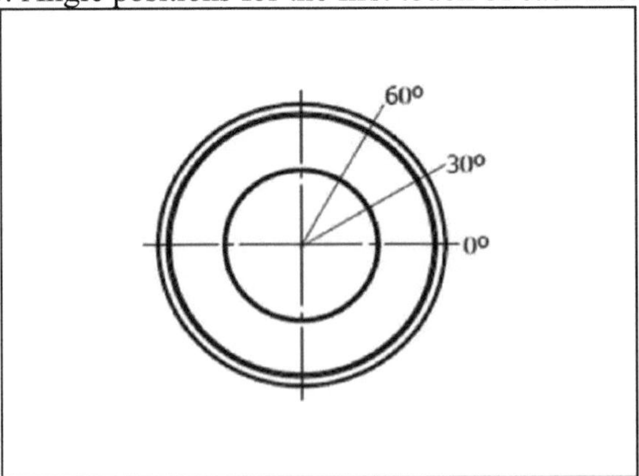

The first diameter was measured by probing four points and then doubling

the number of points until the error stabilized.

The circularity error was considered stable when two average errors obtained in consecutive measurements showed the same value, i.e. the error stabilized in the ninth measurement where 1,24 points were probed, as shown in figure 46.

Figure 46: Stabilization of the circularity error

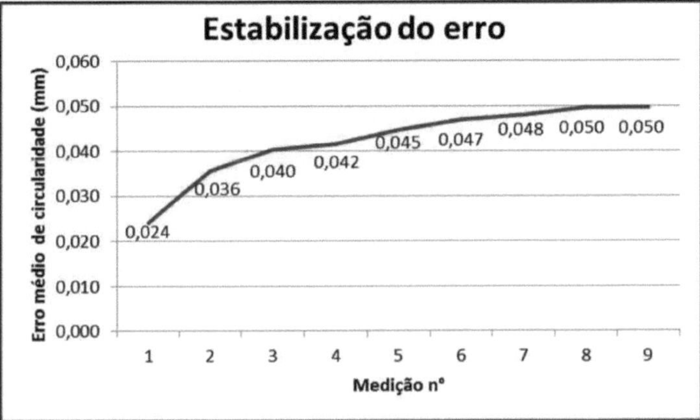

As for the height, the measurements were made in the section that comprises half the length of the cylinder in relation to the measured diameter.

> Measurement procedure

The measuring tip with a diameter of 3 mm was attached to the CMM's measuring sensor and calibrated together with the standard sphere (see figure 47), which has a known diameter, according to the equipment's calibration certificate.

Figure 47: Calibration of the measuring tip

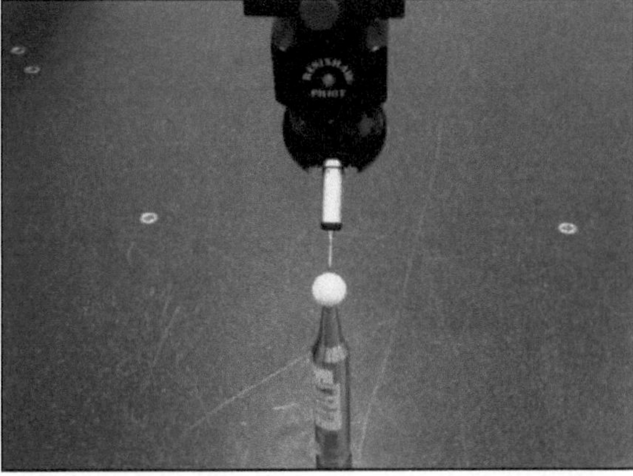

After qualifying the measuring tip, the part to be measured was clamped in the prismatic block positioned on the CMM's measuring table. A plane element was measured with four points on the upper surface of the part and then the plane was leveled on the projection plane formed by the x and y axes.

Next, a circle element was measured at a height of 10 mm from the reference plane and given the origin in the center of the measured diameter, in order to reference the machine for subsequent measurements, which were carried out in CNC mode.

After defining the center of the circle, we started measuring the circles in CNC mode, at the points and angular positions defined in table 4.

Table 4: Number of points and initial angular position

№of points	4	8	16	32	64	128	256	512	1024
	0°	0°	0°	0°	0°	0°	0°	0°	0°
Initial Angle Position	30°	30°	30°	30°	30°	30°	30°	30°	30°
	60°	60°	60°	60°	60°	60°	60°	60°	60°

Measurements were taken for each item in the three predefined angular positions, totaling twenty-seven measurements.

3.2. Current process

The current procedure for measuring circularity error leaves it up to the metrologist to define the number of measurement points, but in general, in CMM measurement with a CNC system, between five hundred and one thousand points are touched for circles with a diameter of up to 100 mm. Diameters above this dimension have their number of points defined according to the functionality of the part in question, and the final decision is usually made on the basis of a consensus between the metrologist and the customer.

3.3. Weaknesses

The main disadvantage of not having a procedure for defining the number of points is the fact that the cost of the measurement may be higher than necessary, since strategies are generally designed with an unnecessarily high number of points to ensure the reliability of the measurement, thus increasing the time spent and consequently the cost.

3.4. Proposal

The minimum number of strokes for measuring circularity will be determined by comparing several measurements.

With the measurement results in hand and by calculating the normalized error, a comparison will be made between the results obtained with the largest number of points and the smallest, one by one, until the minimum number of points with a value of $En < 1$ is obtained.

Once the minimum number of touches has been determined, a relationship will be drawn up between the minimum number of points and the perimeter of the circle, in order to replicate the study for circles with diameters other than the one studied in this paper.

3.5. Results

The measurements were taken for the three angular positions and nine pre-established point numbers. The results obtained can be seen in Table 5, where all the information required for the case study is presented, such as:

- J Number of measurements according to the number of points;
- J Number of points themselves;
- J Values obtained for all angular positions and number of points;
- J Average values obtained;
- J And finally, the measurement uncertainty is calculated for each point, according to the uncertainty sheet.

Table 5: Results obtained

Medigao	№of Points	Value obtained (mm)			Average value obtained (mm)	Temperature (°C)	Measuring time (s)
		0°	30°	60°			
1	4	0,005	0,036	0,031	0,024	20,0	24
2	8	0,039	0,037	0,031	0,036	20,0	42
3	16	0,039	0,041	0,041	0,040	20,0	66
4	32	0,041	0,042	0,042	0,042	20,0	117
5	64	0,045	0,042	0,047	0,045	20,0	231
6	128	0,049	0,045	0,047	0,047	20,0	435
7	256	0,049	0,046	0,049	0,048	20,0	834
8	512	0,051	0,050	0,048	0,050	20,0	1605
9	1024	0,050	0,049	0,050	0,050	20,0	3096

Figure 48 shows the graph constructed from measuring the circle with four touches, starting at the 0° angle position.

Figure 48: Four-touch measurement

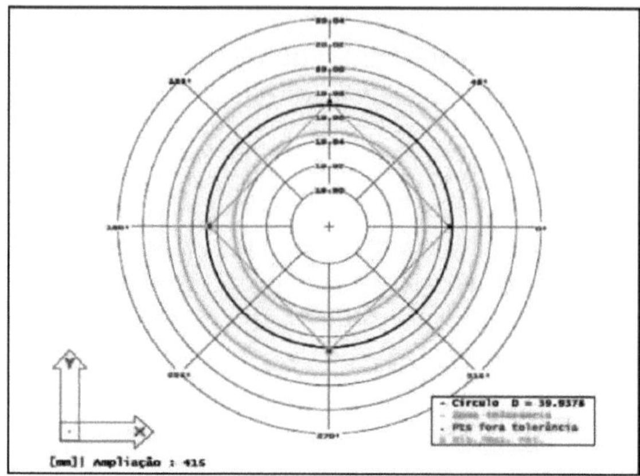

Figure 49 shows the graph constructed by measuring the circle with one thousand twenty-four strokes, starting from the 0° angle position.

Figura 49: Measurement carried out with one thousand twenty-four strokes

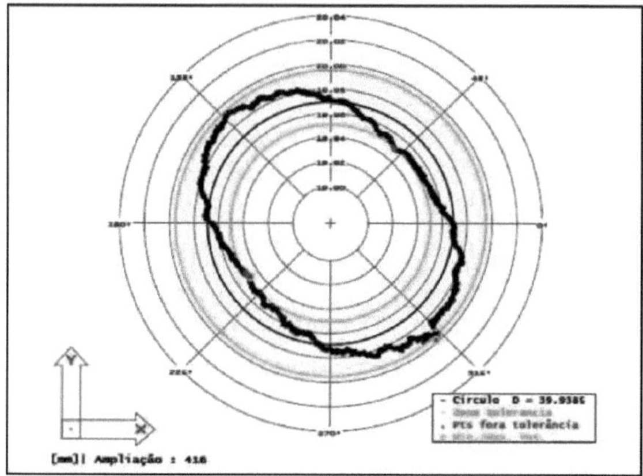

Figure 50 shows the graph constructed by measuring the circle with one hundred and twenty-eight strokes, starting from the 0° angle position.

Figura 50: Medipao performed with one hundred and twenty-eight strokes

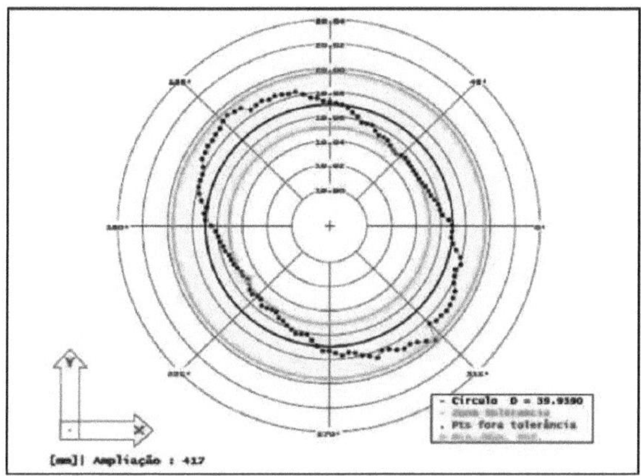

The complete graphs for these and the other measurements can be viewed in the author's files.

50.6. Discussions

Based on the measurements taken, a study was carried out to determine the number of minimum points for measuring a circularity.

> **Step 1 - Calculating the measurement uncertainty:** Based on the results obtained, information on the standards used and environmental conditions, it was possible to determine the measurement uncertainty (U), which is shown in Table 6.

Table 6: Measurement uncertainty (U)

Medigao	№ of Points	Average value obtained (mm)	U (mm)
1	4	0,024	0,019
2	8	0,036	0,005
3	16	0,040	0,003
4	32	0,042	0,002
5	64	0,045	0,004
6	128	0,047	0,003
7	256	0,048	0,003
8	512	0,050	0,003
9	1024	0,050	0,002

The spreadsheet containing the uncertainty components is available in the author's files.

> **Step 2 - Calculating the normalized error: Once the** results and measurement uncertainties had been obtained, it was possible to determine

the normalized error by taking as a reference the measurement with the largest number of points, which in this study's case is the ninth measurement with 1,24 points. Table 7 shows the normalized errors.

Table 7: Normalized error

№	Comparing	Result24		Results	MORE	E.N.
		Parameters	U-1024	Parameters	TOO MUCH	
1	(1024e4) points	0,050	0,002	0,024	0,019	1,36
2	(1024 and 8) points	0,050	0,002	0,036	0,005	2,60
3	(1024 and 16) points	0,050	0,002	0,040	0,003	2,77
4	(1024 and 32) points	0,050	0,002	0,042	0,002	2,83
5	(1024e64) points	0,050	0,002	0,045	0,004	1,12
6	(1024 and 128) points	0,050	0,002	0,047	0,003	**0,83**
7	(1024 and 256) points	0,050	0,002	0,048	0,003	**0,55**
8	(1024 and 512) points	0,050	0,002	0,050	0,003	**0,00**

> **Step 3 - Graphical analysis of the normalized errors:** The graph shown in figure 51 demonstrates the variation between the measurements, where the distinction between acceptable and unacceptable values is made by means of the colors of the bars. The blue bars represent acceptable values (En < 1) and the red bars represent unacceptable values (En > 1).

Figure 51: Graph of normalized errors

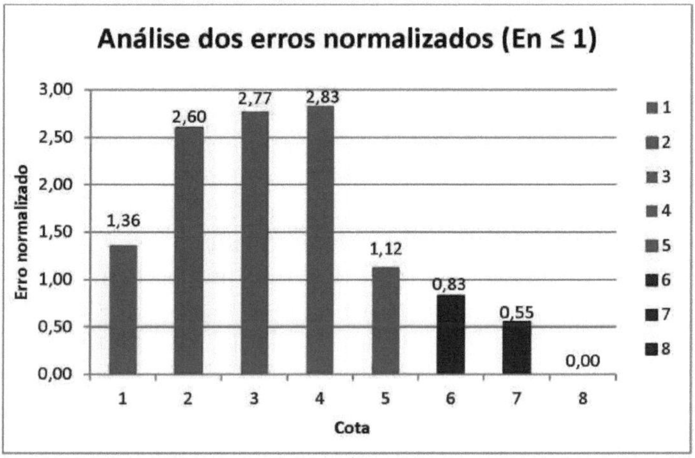

From the analysis of the normalized errors, it can be seen that the best number of points for measuring the diameter of the pepa in this study is 128, since it is the smallest number of points that showed En < 1.

> **Step 4 - Defining the number of points according to the perimeter of the circle to be measured:** Knowing that the sixth measurement represents the best option (in terms of productivity with quality) for measuring a diameter of 40 mm, a rule was established for defining the number of points to be touched in any circularity measurement according to the perimeter of the circle to be measured. Equation 8 determines the calculation of the perimeter.

Where:

$$P = 2 \, . \, \pi \, . \, r \tag{8}$$

P - perimeter;

r - radius.

The nominal diameter of the circle measured for this study is 40 mm, so the radius is 20 mm. Substituting into equation 8:

$$P = 2 \, . \, \pi \, . \, 20$$

$$P = 125{,}664 \, mm$$

With the perimeter you can determine the number of stitches / mm:

$$\frac{128}{125{,}664} = 1{,}018$$

In other words, approximately 1 point/mm of the calculated perimeter.

Therefore, the rule is to determine the perimeter by means of the nominal diameter and from this information establish the number of points to be touched considering the ratio: 1 point/mm of the calculated perimeter.

1.7. Advantages of using the proposed rule

One of the advantages of using the rule proposed in this work is the increased reliability of the measurement, since there is a standardization for determining the number of points in circularity measurements at different diameters.

However, the main advantage becomes clear when analyzing the reduction in the time needed to measure a circularity. Using data obtained in this study as an example, it can be seen that for measuring a diameter of 40 mm, the rule establishes a number of 128 points, taking 435 seconds. Before using the suggested rule, it was estimated that the number of points required to measure a pepa with the same diameter would be between 500 and 1000, so the (experienced) number of points closest to the minimum number was taken, which in this case is 512 points with a time taken of 1605 s. An analysis of the variables shows that using the rule provides a time gain of at least 72.9%.

Conclusion

The study carried out in this paper proved how easy and fast it is to obtain shape errors using coordinate measuring technology. This technology, which has been developed since the invention of the CMM, together with the application of GD&T concepts, brings great benefits to production systems, guaranteeing productivity with quality.

The execution of the case study led to the development of a rule that allows for greater reliability in the measurement, as it establishes a standardization for circularity measurements in different diameters.

In addition to the advantage gained from the increased reliability, a time saving of around 72.9% could be seen when comparing the techniques for defining the number of points used before and after creating the rule.

Therefore, it can be concluded that the rule consisting of touching 1 point/mm of the perimeter of the circle to be measured provides higher quality and lower product development times and costs.

It's worth remembering that the work shown here was only carried out on circularity, so here are some suggestions for studying other types of shape errors in the future: cylindricity, straightness and flatness.

References

ABNT / INMETRO. **Guide to the Expression of Measurement Uncertainty:** Third Brazilian edition in Portuguese. Rio de Janeiro, RJ: ABNT, INMETRO, 2003.

ALBERTAZZI, Armando; SOUSA, André R. de. **Fundamentals of Scientific and Industrial Metrology**. Barueri, SP: Editora Manole, 2008.

AMERICAN SOCIETY OF MECHANICAL ENGINEERS. **ASME Y14.5:** Dimensioning and Tolerancing - Engineering Drawing and Related Documentation Practices. New York: ASME, 2009.

ASSOCIAÇÃO BRASILEIRA DE NORMAS TÉCNICAS. **NBR 6409:** Geometric tolerances - Shape, orientation, position and strike tolerances - General, symbols, definitions and indications in drawing. Rio de Janeiro: ABNT, 1997.

BOSCH, J.A. **Coordinate Measuring Machines and Systems**. Marcel Dekker, Inc., 1995.

DIAS, José Luciano de Mattos. **Measurement, Standardization and Quality;**
Aspects of the history of metrology in Brazil. Rio de Janeiro, 1998.

EURACHEM / CITAC. **QUAM: 2002 - Brazilian version of the Guide**
EURACHEM / CITAC: Determining Uncertainty in Analytical Measurement. Rio de Janeiro: QUAM, 2002.

INMETRO. **DOC-CGCRE-005:** Guidelines for the Organization of Interlaboratory Comparisons by Dicla Technical Commissions. Duque de Caxias, RJ: DOC-CGCRE-005, 2002.

INMETRO. **NIE-CGCRE-009:** Use of the Mark, symbol and references to accreditation. Duque de Caxias, RJ: NIE-CGCRE-009, 2012.

INTERNATIONAL ORGANIZATION FOR STANDARDIZATION. **ISO 1101:** Geometrical Product Specifications (GPS) - Geometrical tolerancing - Tolerances of form, orientation, location and run-out. Geneva: ISO 1101, 2004.

KLEN, Edmilson Rampazzo. **Calibration of Coordinate Measuring Machines Using Ball Plates - An Economical, Fast and Metrologically Efficient Technique.** Florianópolis, 2000. Master's dissertation in Scientific and Industrial Metrology, Federal University of Santa Catarina.

LIRA, Francisco Adval de. **Metrologia na Indústria**. Sao Paulo: Editora Érica, 2009.

MITUTOYO, Metrology Institute (Training Department). **GD&T:**

Dimensioning and Geometric Tolerances. Sao Paulo: Mitutoyo, [200-].
MITUTOYO. ***Geopak-Win*** **- 3D Geometric Measurement Program for Coordinate Measuring Machines.** Suzano, SP: MITUTOYO, 2003.
MITUTOYO. Products - **CRYSTA-APEX Coordinate Measuring Machines**
C. Brazil, 2012. Available at:
http://www.mitutoyo.com.br/produtos/index.html
Accessed on: 03-Oct-2012.
PASTORE, Julio Cezar. **Geometric Shape and Position Tolerances**. Jundiaí, SP: JP Veritas, [200-].
PIEKAR, Daniel Samuel Hamburg. **Calibration of Standard Peps in Coordinate Measuring Machines.** Florianópolis, 2006. Dissertation Master's Degree in Scientific and Industrial Metrology, Federal University of Santa Catarina.
PORATH, Maurício de Campo; SOUSA, André R. de. **The Technology of**
Coordinate Measurement in the Product Development Cycle Plastics. Florianópolis, SC: PORATH e SOUSA, [200-].
NATIONAL INDUSTRIAL LEARNING SERVICE. **GD&T:** Dimensioning and Geometric Tolerancing. Rio Grande do Sul: SENAI, 2007.
STRAFACCI NETO, Gilberto. **GD&T: Why?** Sao Paulo: Setec, [200-]. Available at:
<http://www.setecnet.com.br/qualidade/includes/GD&T_Porque.pdf>
Accessed on: 13-Jun-2012.
VIM, International Vocabulary of Metrology: **fundamental and general concepts of associated terms (VIM 2012)**. Duque de Caxias, RJ: INMETRO, 2012.
WANDECK, Maurício. **GD&T On Line**. GD&T Engenharia Ltda. Rio de Janeiro, RJ: WANDECK, 2010.

Printed by Books on Demand GmbH, Norderstedt / Germany